HOW TO SUCCEED AS A GIS REBEL

A JOURNEY TO OPEN SOURCE GIS

MARK SEIBEL

LOCATE
PRESS

Credits & Copyright

How to Succeed as a GIS Rebel

A Journey to Open Source GIS

by Mark Seibel

Published by Locate Press

Editor Tyler Mitchell
Interior Design Based on Memoir-LATEXdocument class
Cover Design Nathan Watson & Julie Springer
Publisher Website http://locatepress.com
Book Website http://locatepress.com/book/osgis

Version: 01d55ad (2023-09-07)

Contents

Preface

Welcome to the first of a new kind of book for Locate Press, one that is less about the "how to" of technology and more about navigating the professional GIS landscape and its application.

Meet Mark Seibel, this book shares the lessons he learned throughout his 25-year career as a GIS professional. The most notable transition during that time was his move to using open source software as a core feature of his geospatial workflow. How did he get there? Read on to learn about the successful, though at times frustrating, journey through education, trial and error, and how he made himself indespensible at his work without breaking the bank.

This story is near to my heart as Mark's journey aligns with my own, and many others, who went through the early days of the open source movement. Many exciting things spun out of early OSGIS journeys and I know that your path can be just as impactful and amazing.

If you have ever wondered how you take what you already know, mix in some open source technology, and increase the payback to your professional life, you'll enjoy following along. This content is a sort of "GIS in a nutshell"—helpful for those considering GIS as a career but also for experienced professionals trying to see the bigger picture.

Naturally, we expect more people will continue to adopt open source concepts that encourage working together and playing well with others. It's our hope that this book helps build that case for you. As always, thank you for supporting our work and keep it touch.[1]

Tyler Mitchell
Publisher

---

[1]Sign up for news and exclusive discounts: *https://locatepress.com/subscribe*

*To my beloved wife, who has always believed in me
and encouraged me to follow my dreams.*

"The journey is the reward."
Chinese Proverb

1. Introduction

What is open source GIS software and what can it do? This book aims to answer these and other related questions told through a 25-year journey using Geographic Information System (GIS) technology. This journey started in 1995, learning about commercial GIS in university on the UNIX operating system. It ends in 2022, exclusively using open source GIS software on Linux as a freelancer. Many desktop GIS applications were encountered, as well as spatial databases and web mapping platforms. The transition away from expensive GIS software took some effort and one aim of this book is to share this information through a collection of experiences.

1.1 Who is this book for?

Have you ever wondered what GIS is? Maybe beyond that, who uses it and what can it do? Maybe you are in college trying to figure out a path or maybe you have already started your GIS career and wonder what other areas of GIS might be interesting. Maybe you're one of the people who are changing their profession to GIS. It's possible that you have used GIS in the past and are now looking to take it to the next level. Maybe you're interested in the open source component of GIS. Others still may not have a genuine interest in GIS but are required to use some GIS software to pre or post process data for their needs. There are certainly some who are not directly using GIS software but are in a position that they need to know about it and manage people who are using the technology.

Regardless of why someone may have an interest in GIS, this book aims to explain open source GIS software and applications through a 25-year personal journey. There's no hidden agenda—our intent is to help bring more people into the open source community.

Have you ever considered getting into the GIS profession or just using the software in some capacity secondary to your primary duties? GIS software has such high visibility that it seems unlikely people wouldn't have at least heard about it in academia or in their career by now. GIS users range from full time users to partial users. By partial we mean they have a primary profession that is not GIS, but they need to use GIS software as part of their main job function. For partial or casual users, GIS isn't really a profession so much as

1

a tool the professional uses to accomplish their job requirements. An example of this would be a wildlife ecologist mapping some listed species.

As an analogy, there are some software users who can create a word processing document or edit some graphics. On the other hand, there are writers and artists who can use this software in a professional capacity for their career. In the same way, one can dedicate their career to using GIS software and related technologies. However, dedicated GIS users are more than just software users. They have a deeper understanding about GIS and knowledge about how it integrates with other technology. They also have an in-depth understanding about GIS data, which becomes as important to know about as the software. We'll dive into data details later.

GIS is more than making maps or even analysis. The technology has evolved into a broader system than its initial conception, mostly due to the Internet. As GIS expanded, more people could access it through web based mapping or other technologies such as an enterprise-based Relational Database Management Systems (RDBMS). GIS desktop software, relational databases, web technology and mobile data collection all work together to provide an Enterprise GIS—a term that has emerged describing a robust system that can store and distribute GIS data to an organization.

1.2   Off we go

If you are new to GIS, it's important to think in terms of choosing a starting point, not a path. The path will change and evolve over time as your knowledge, experience and interests expand. There are many areas within GIS and as your interests change over time GIS can easily accommodate many of these personal changes. If you are already using GIS on a daily basis, know that there is always something new to do in GIS if you have the passion and interest to explore it.

The roles in this personal GIS journey run the gamut from: GPS technician, Senior GIS Analyst/Programmer, GIS/CAD Manager, GIS Technical Lead, Spatial Database Administrator, Web Administrator, Linux System Administrator, GIS Web Administrator (GeoServer, MapServer, GeoNode), Senior Data Engineer and finally, author. This journey was trail-blazed with the help of the open source community and entailed a lot of trial and error to solve real world problems for major corporations. There was documentation for open source GIS in the early days, but it was not as digestible and plentiful as it is today. [One of the primary reasons that we started Locate Press! —Ed.]

Some tasks in GIS amount to sequential button pressing, while other tasks are more complex, requiring deep thinking and methodical approaches to be deployed. It's important to figure out how you enjoy using GIS, so that the work you do will be fun and interesting each day. Do you have a knack for colors and symbols and enjoy creating aesthetically pleasing maps? Or do you get excited about modeling features and crunching big data? Perhaps systems and web mapping are interesting to you, or you enjoy administering enterprise database systems. What about leading a GIS team? There are so many ways to be involved with GIS and the best part is you get to choose how.

1.3 Secret's out

The secret is out: GIS is fun. Let's face it, not many people look at CAD, get excited, and want to tinker with it in the same way they do with GIS software. GIS software has a unique appeal to many people. Due to its popularity and user growth, the GIS field has become very overwhelming and often unclear for new prospects to determine where to start and what skills are important. This book aims to help guide those in this situation. New users may be students, current professionals, or the curious-minded. Open source software should definitely be one of these curiosities explored as it's still a bit of a secret.

2. What is GIS

2.1 Dizzying GIS Definitions

Before we get to the good stuff, we should define some terms, two in particular: *open source* and *GIS*. We'll start with GIS and explore open source and proprietary software in the next chapter. The question "What is GIS?" can be answered with one word: software. Remember this simple answer. GIS is just software, like a word processor is software. It's not special, it's just spatial.

It can be a hard pill to swallow because most people who chose GIS as their profession don't want to admit they're just software users. On the plus side, once it's realized, it can open a new way to thinking about desktop GIS, supporting infrastructure, and where the most fun is to be had in the GIS realm. If you have a knack for software you're in for a real treat. If you are also into Linux, you are in for an *absolute* feast.

As formalities go, there are many definitions for a Geographic Information System (GIS). Summaries are presented below from DJ Macguire's "Overview and Definition of GIS" (Macguire, 2013). As you will see, there are commonalities among the definitions, but subtle distinctions are called out:

"DOE (1987:2132) - a system for capturing, storing, checking, manipulating, analysing and displaying data which are spatially referenced to the Earth.

Aronoff (1989:39) - any manual or computer based set of procedures used to store and manipulate geographically referenced data.

Carter (1989:13) - an institutional entity, reflecting an organizational structure that integrates technology with a database, expertise and continuing financial support over time.

Parker (1988:1547) - an information technology which stores, analyses, and displays both spatial and non-spatial data.

Dueker (1979:1061) - a special case of information systems where the database consists of observations on spatially distributed features, activities, or events, which are definable in space as points, lines, or areas. A GIS manipulates data about these points, lines, and areas to retrieve data for ad hoc queries and analyses.

Smith et al. (1987:13) - a database system in which most of the data are spatially indexed, and upon which a set of procedures operated in order to answer queries about spatial entities in the database.

Ozemoy, Smith and Sicherman (1981:292) - an automated set of functions that provides professionals with advanced capabilities for the storage, retrieval, manipulation, and display of geographically located data.

Burrough (1986:6) - a powerful set of tools for collecting, storing, retrieving at will, transforming and displaying 3 spatial data from the real world.

Cowen (1988:1554) - a decision support system involving the integration of spatially referenced data in a problem-solving environment.

Koshkariov, Tikunov and Trofimov (1989:259) - a system with advanced geo-modeling capabilities.

Devine and Field (1986: 18) - a form of MIS [Management Information System] that allows map display of the general information."

—Macguire, 2013

The DOE definition is succinct and hits on the key points: capture, store, check, manipulate, analyze and display geographic data (which is data referenced to the Earth):

Capturing: Digitizing geometry (point, line, and polygon) that represent features in the real world.

Storing: Refers to where the data lives, such as on a local computer, network, or cloud configuration.

Checking: This can be vague, but could refer to querying the data, such as asking the system "how many acres of wetland are present in this parcel?"

Manipulating: Editing data, which is a straightforward process with exception for data topology management.

Analyzing: Functions performed on the data to derive new outputs such as combining layers, buffering features, or performing statistical summaries.

Displaying: What good are all of these capabilities if the awesome data outputs cannot be viewed? So, a GIS needs to be able to display or illustrate the data, also known as cartography: the art and science of making maps.

2.2 Expanded Definition

Previously, we simplified GIS to a one word definition: software. While true, there obviously needs to be elaboration on the kind of software. Maybe starting with the data helps explain what the software needs to do.

GIS works with spatial data, which is data that has a reference to a location on Earth. GIS features are a point, line, or polygon with coordinates. These features have attributes and the attributes are stored in a database or table. This is how we click on a feature and get information about it. One example is a railroad line that has attributes describing what kind of rails, who owns it, and the last time the track maintenance was performed.

Expanding our definition further, GIS is an applied technology to the sciences. Applied technology must have broad-based use and flexibility, not just perform a single task like how to get from point A to B along a road network. It's an applied technology because a GIS by itself is useless software without some data to work with. The data is usually related to a scientific field or topic and has to follow those constraints. An example would be mapping wetlands, modeling where water flows, or mapping a transportation network. A word processor needs data to be useful and the same is true for other software. GIS started with environmental applications, so its extension into other sciences is a natural fit. Nearly every feature on Earth can be modeled as a point, line, or polygon; and if not, we have raster 2.5D elevation models and true 3D voxel data.

What is a GIS? At a basic level a GIS makes maps, but we're interested in much more than that. It's a system that allows questions to be answered. Not just in the typical SQL fashion, "How many customers do we have?" but more complex questions that involve geographic areas. An example of a somewhat simple problem for GIS to solve is marketing beach gear for children. "How many customers do we have in the $150K-$200K household income range, with 2+ kids that live near the beach?"

A more complex example would be siting a solar power plant. The desired areas must meet these criteria:

1. large agriculture fields or pastures greater than 100 acres
2. proximity to a transmission line (max of 2000 feet)
3. the furthest distance from wetlands, water bodies and streams,
4. the furthest from residential zones
5. closest proximity to industrial or commercial zones

It's not just an information system that also queries data and answers questions, it's a *geographic* information system that can integrate all kinds of interesting location components into the query.

As stated in the opening definition, the GIS must be flexible; it must possess tools that allow a user to express their thoughts through geographic functions. This really becomes the key to unique applications of GIS software. With this understanding, the GIS software needs to be able to do more than one task or provide more than a single service. It must be flexible in how it can operate on different GIS data so the operator can solve a wide range of problems. It needs to be able to determine how many acres are impacted by development just as easily as finding an optimal location for a new electric vehicle charging station. What's realized over time is that the functions and tools to solve spatial problems are consistent but the problems keep changing.

First pop quiz: Is Google Maps a GIS?

No, it's an online mapping service. Google Maps is awesome and it can do a lot of cool things. It can find the fastest route from point A to point B with accurate time estimates which include using live traffic data. It has an enormous inventory of points of interest and it also can overlay aerial photographs as the base layer. It may seem to meet the simplified definition for GIS on the surface, but it lacks the ability to perform open-ended analysis.

2.3   What Can GIS Do?

All these definitions and attempts to describe GIS are fine, but the best way to describe GIS is to give examples about what it can do. From an environmental perspective, it can map and summarize wetland impacts on a development project. It can use satellite data to determine how land cover vegetation types change over time, such as identifying deforestation in rainforests. GIS can also determine the volume of earth for cut and fill analysis. Using elevation data, hydrologic modeling is possible, determining where water flows and how much accumulates in depressional areas. GIS can find suitable land areas meeting specific criteria, such as finding the best location for a solar power plant, identifying constraints and non-buildable areas. These are just a few examples, the list is as long as the imagination is creative.

Spatial Functions

GIS software offers spatial functions that can create outputs from one to many GIS layer inputs.

The simplest example of a geometric function is the *buffer*. This can take a feature and apply a polygon offset from the shape to a specified distance. This is useful for determining proximity from existing features.

Another function is the *clip* function. It's like a cookie cutter in that cookie dough is the data and an area of interest is the cookie cutter. The input dough is clipped to the bounds of the cookie cutter.

Other functions include overlays where layers are combined - such as a *union* or *intersect* - to create new geometry and attribute combinations. This is useful for finding commonality between layers, or the reverse, exclusivity. If we buffer a listed species point 2,000 feet, we can union this layer with a wetlands layer to see where the listed species buffer is in wetlands.

The last of the basic functions is the *erase* function. This tool which deletes features in one layer with features in another layer.

Knowing how to press these buttons is very useful, but the one thing that begins to separate levels of GIS users is the ability to understand how these functions can be used in a larger, more complex analysis. All of these functions can be, and often are, chained together in a pipeline that drives towards a specific output. Beyond that, sometimes these tools can be useful outside what is obvious.

2.4 Why use GIS?

Now equipped with a basic idea about what a GIS is and what it can do, the next question is why use a GIS? At a basic level the answer is "to make maps". Beyond that though, a GIS is a database; so a more helpful answer is, "GIS answers questions".

Oftentimes, GIS is simplified to a map and the true power of GIS software is not apparent to most people at first glance. It's different from database queries in that there can also be spatial queries based on the shapes of features and where they are located on Earth.

What a GIS can do is limited to what data is available. So, to be really good with GIS, one must know about, and being able to use with complex interaction, a wide variety of data. Some examples of GIS data are: road lines, telecommunication networks, land use areas, streams, wetlands, satellite imagery, LiDAR point data, digital elevation models, parcel ownership and facility locations. Since these data can have spatial relationships with each other, GIS can integrate the different data and begin to process the data accordingly. The combination of a very knowledgable and skilled GIS user can really do some amazing things.

The good news is that a lot of GIS data is freely available. Provided the data exists, one could use GIS to answer these types of questions:

- What are the land use categories for these land parcels?
- How many people are within a 1/4 mile of public transit stops?
- How many acres of wetlands will be impacted by the development of this land?
- What are the shortest and fastest routes to get from point "A" to points "X, Y and Z" on this road network?
- Where does the water flow on this large tract of land?
- Where is the best place to put a new convenience store?
- How did the vegetation cover change in this area over decades?
- If this cell tower is erected, what homes will be able to see it?
- Where is the optimal place to build a solar power plant?

2.5   Most Important Component in GIS

What is the most important component in a GIS? It's a bit of a trick question. Interestingly, it wasn't mentioned in any of the definitions above. Earlier we said GIS is software, but the thing with software is that it must have users. This brings us to the most important component in GIS: **users**.

Software is just code that sits and waits for people to tell it to do something. In the case of GIS, this is even more pronounced by the vast number of tools that can be combined in a variety of ways to create methodologies for solving complex problems. The GIS user is the key to making GIS software do incredible things, just like any software really.

Note that the prior definitions of GIS, both formal and informal, exclude the most important component of a GIS: the person operating the software. This exclusion is intentional because it reinforces the idea that GIS is software. Software just exists but people make it do special, in our case, spatial things. Writers make good books, not a word processor. The same is true for spreadsheets and engineers, and graphics software and artists. For truly remarkable outputs, the value is always in the user, not the software.

GIS definitions do not include the operator and they shouldn't because Information Systems are *machines* not *people*. The operator is the one who can really unleash the power of GIS depending on their ability to problem solve. This is because people have the ability to use logic and think around corners, while machines can't do that yet. People have the ability to think in abstract ways and develop innovative processes to solve difficult problems. Some examples will be illustrated in detail during the journey in this book.

Runner up for the most important component in a GIS is: **data**. Data is definitely a close second place because GIS can do nothing without it. The same can be said for GIS operators, but operators take first place because of their analytical capabilities; and they're people. Understanding the data is just as important as knowing how to run the software. GIS data comes in many formats, from many time periods and with varying resolutions. Knowing how to combine and use these effectively is an essential bit for real growth with GIS software and solving modern day problems.

2.6 GIS Titles and Roles

What do you do?

As a broad-based GIS user, one of the difficult questions to answer is, "So what do you do?". There begins the problem of trying to describe what one does with GIS for a profession to a non-GIS or even non-technical person. Here enters the labeling of people who use GIS software.

For the first five naive and simple years using GIS in a professional setting, it was difficult to give a concise answer to the question "So, what do you do?". Responding with "I am a GIS Analyst; I work with Geographic Information Systems or GIS." would only produce a blank stare in return. Proceeding to elaborate on GIS at the nerd level would quickly turn the blank stares into tears as the face of utmost regret showed itself.

Based on the reactions, it was evident that some other way to describe the work was needed for people who did not know about GIS. Remember, this was before Google Earth which came out in 2005, so there weren't quick answers available that people could easily relate to other than maps. During the early 2000s, GIS web services were evolving, and at the time MapQuest was a relatable GIS web service. Interestingly, MapQuest fell off the Internet after Google removed it from top search results in 2007. By 2005, we had Google Earth, a relatable software product to people—minus key features—when trying to provide a GIS software analogy.

Over time, the reply to the question "What do you do?" evolved into a response of "computerized mapping". A small example would follow, such as determining the impacts and enhancements of mining activities to wetlands and streams. People seemed to understand this, but the simplification didn't mention all the other aspects being integrated like automated mapping through programming and GPS field data integration. This response was suitable in the early years when things were simple. After much growth and experience using GIS software for complex analysis and modeling, this response was not expressing the full value or capabilities of how GIS was being used.

Sometime later in the journey was a discussion with the IT Director who referred to the GIS work as "environmental modeling". This was very good. The answer related to environmental science and also computer modeling. The sound of this was appealing because it was more specific, relatable to people, and distanced itself from the simplistic cartography aspect of GIS. This was really a better description of the specific work being done. However, it was too specific to the environmental field and excluded other GIS technology that was being implemented with open source software.

By the late stage of the journey, the answer just evolved into "I'm a computer nerd". After so many years of highly technical GIS work on the desktop and with servers, it's the best answer. Most people seem content with this nerdy answer and return the answer with a smile rather than a blank stare or even tears. If people want to know more details about the work, details are provided. The open source software, both geospatial and non-geospatial, is definitely operating at the nerd level.

Formal Titles and Informal Roles

Organizations have formal titles for GIS users and there are also informal roles that users may assume. Over a long period of time, it was interesting to watch the GIS technology evolve and the titles that emerged with it. The titles fit different GIS activities such as desktop GIS user, GIS web administrator, or programmer.

In the 90s, there wasn't really much more to GIS titles than GIS Technician, Specialist, Analyst/Programmer and GIS Manager. The progammer part of programmer/analyst was more about scripting initially. There weren't GIS web developers, GIS software developers, or online GIS systems to administer. Sure, there were GIS developers because GIS software was being made, but there weren't mainstream GIS developers who could build their own GIS from a set of programmable tools.

Unfortunately, open source had not been encountered in this journey to know how users were developing GIS software in the open source community. There is no doubt this was happening at the time, based on evidence of historic activity and then the future software encountered, but it was not on the radar so far in the journey.

Labels for GIS users fall into the two categories of roles and titles. Titles are defined by industry as official job positions, but roles are informal and usually an add-on to the worker's existing title. As far as titles go, GIS has become interchangeable with the word geospatial and thus some of the titles in modern day reflect this. For example, it may be Geospatial Technical Lead rather than GIS Technical Lead. Geospatial is a broader field which encompasses GIS and other locational technologies such as Global Positioning Systems (GPS). GPS is a key component for field data collection.

GIS Titles

GIS positions can have varying responsibilities, but generally follow these guidelines. The titles are specific to users who primarily use GIS in daily activities. GIS titles are formal names for positions often seen advertised as jobs.

GIS Technician - Entry-level GIS operator: Basic skills include creating simple point, line and polygon data. Should be able to create functional maps with existing GIS data. Minor table experience in sorting and querying. This is the beginning level where both professionals and basic GIS software users posses roughly the same skills.

GIS Specialist - Mid-level GIS operator: Possesses skills of the GIS Technician plus additional skills that are gained as experience and knowledge grow. Some examples include topology management of vector data or gaining experience with raster data, or performing basic data overlays and statistical tables. There's an interesting transition opportunity here because progression beyond this level requires analytical capabilities; or at least should.

GIS Analyst - Top-level GIS operator: Possesses skills of the GIS Specialist plus functional analytical capabilities. This level should be far removed from the early days of cartography and simple data creation, and far more aligned with problem solving. Absorbing real world problems and knowing not only which tools to use, but the order to use them. Experience and knowledge which has led to the ability to work with multiple data types. Thorough understanding about data types and applications.

GIS Data Librarian (Steward): Person responsible for GIS data maintenance. Updates base layers used by the organization while maintaining quality data and attributes.

GIS Data Scientist: Deep dive into data using tools for statistics and advanced data slicing. Develop processes and methods for manipulating and analyzing data. Examples could be working with LiDAR data in the crop industry and scientific data based on crop health or pests.

GIS Manager: At a minimum, possesses basic GIS skills of the technician as a GIS software operator, but generally does not perform GIS desktop work. Primary value added is managing or leading a group of GIS workers, but also coordinating work flow, obtaining external work for the group and working with people across the organization to integrate GIS with their work efforts.

GIS Software Developer: A programmer dedicated to GIS software development. Often building tools or using components of a GIS to build specialized applications. Advanced GIS often requires writing scripts to automate repeated processes. It's worth noting that writing scripts is great and awesome, but is not the same thing as an application developer who can create stand alone applications to run on various platforms.

GIS Architect: Primary job duty is to put all the pieces of GIS together in an organization to provide enterprise solutions. GIS is more than operators using desktop software; there are server components, web-based technology and other ways to integrate GIS into an enterprise's workflow. The architect identifies various software pieces that enable GIS functions to be utilized across the entire organization, but their role diminishes once the system has all the desired infrastructure.

GIS Administrator: This person's responsibility is orientated around managing GIS software; desktop and server applications. Note crossover with IT functions. They may be in charge of a large centralized GIS data library or be in charge of keeping desktop software up to date. In a large organization, this may be a stand-alone role, but in smaller organizations it is usually an added responsibility of a GIS Operator role.

GIS Roles

GIS roles are informal positions, often an auxiliary role to a primary position. Usually an add-on responsibilities or necessary usage of the software. The roles below are individuals who are dedicated to the profession.

GIS User: This is just someone who uses GIS software. Just like they would use a spreadsheet or word processor, it's just a tool to work with specific data, geographic data. This may be a scientist who uses GIS to explore geographic data. It could be an engineer who uses GIS to prepare data for water resource modeling. It's hard for GIS users to become experts in GIS because their time is dedicated to their primary field of expertise. Sure, the computer nerds defy this statement.

GIS (Geospatial) Technical Lead: Technical leads have become well established in organizations. A technical lead provides guidance and shares their expertise with the intent to lead others down a successful path. They also play a role in leading the organization down a sustainable path for GIS. The difference between a manager and a leader is another topic, but probably the best distinction between a manager and a leader was found in a fortune cookie: "Managers do things right, while leaders do the right things."

2.7   GIS Professional Certificate

In addition to these titles and roles, there is a formal *GIS Professional* (GISP) certificate that can be obtained once specific criteria has been met by a governing body called the GIS Certification Institute[2]. *The GIS Certification Institute* validates education, experience and contributions to the profession for each GISP. This is further discussed in a later chapter, but worth noting it's the only credential or certification GIS Professionals can get at this time.

2.8   Two User Paths

Direct Role

You're either a primary user of the software as your area of expertise or you use GIS technology to help with your primary area of expertise. This is the main division of GIS software users.

---

[2]https://www.gisci.org

It may not be clear at first, but there are two paths to consider with regards to GIS software. The first path is a direct, constant usage of GIS software, applying GIS to various applications and industries.

The direct role is your geonerd; your dedicated GIS professional who knows the software extensively through daily usage and the concepts that apply to the data and applications. They use GIS software as their primary job function, working in various professions and disciplines. The direct role would include creating data, performing analysis and making maps. A direct role with GIS software is really a fun proposition because GIS is an applied technology to all the other wonderful sciences.

Indirect Role

The indirect role is still a direct interaction with the GIS software, but GIS software is a secondary skill to a primary profession such as a scientist, engineer, developer or manager. For the non-direct roles, GIS is a secondary ability to their primary area of expertise. One example would be a water resource engineer performing floodplain modeling. Floodplain modeling takes into account a lot of things like elevation, storage depressions, drainage networks, drainage areas, soils and surface roughness. All of these features are usually in a GIS format and used as input for the floodplain modeling. The floodplain modeling is not necessarily done in GIS, so GIS is considered a data preprocessor in this example. The surface water engineers get very experienced with these GIS components of a floodplain model, but beyond that their knowledge in GIS can be quite shallow.

In a real-world example, we might have a web developer, software developer or database administrator who has expertise in each of those areas, but needs to incorporate GIS components into their infrastructure. In this example, the GIS component is secondary to the primary technology of web, software and database development. Non-direct roles do not require a primary expertise in GIS. There is only so much time a person can spend in their profession to become an expert. One can't satisfy both the primary expertise of GIS with another primary area of expertise such as ecology or planning.

2.9 Careers using GIS

There are many career professionals that use GIS software, but they fall into the two categories described above: GIS professional and GIS software user.

We already touched on this one, but the obvious career in GIS is a GIS professional who uses GIS software on a daily basis; a dedicated user. This is where most hardcore GIS users end up. Technician, specialist, analyst or other title, they're all positions that interact with GIS on different levels on a daily basis. This is a fun and interesting career choice, but always be aware that with changing technology you'll eventually have to adapt to different software and possibly workflows.

Educators are the ones sharing and spreading the GIS knowledge. These are our professors and teachers sharing GIS concepts with students and helping them gain a solid foundation. Teaching GIS can be a rewarding career, especially when done in a way that doesn't limit student's thinking to only consider commercial GIS software.

> The term "commercial software" in this context can be interchanged with "proprietary software", as software that is not free to use and without publicly accessible code. Commercial is used throughout this book to help highlight the fact that it involves a high price point to join and undergirds the traditional GIS market as a mature, and expensive, product offering. Meanwhile, "proprietary" is more about software licensing and code sharing and less about the business approach. More on this in the next chapter.

GIS developers write a lot of code to make GIS applications. These applications can be user facing directly as software, or more subtly operating in the background with back-end software. In the open source world, developers play a special role being GIS software users in addition to developers, a benefit commercial software can't partake in just by the nature of their environment. In a commercial environment, the developers may not have spatial backgrounds or even interest, which can negatively impact the GIS software.

Marketers and recruiters are trying to sell GIS software or GIS users. These individuals can be well versed in GIS but often can't do much of anything with it—managers can fall into this camp too. Sometimes these roles can be a transition after having had hands-on experience with GIS. These positions require some extraversion and the special ability to work with people more than machines.

Geospatial Architects look at various components of spatial technology and implement geospatial software pieces that service an organization or enterprise. A geospatial architect works with developers to build code specific to an organization's needs. Examples would be setting up a web server to distribute GIS data and information, or designing a centralized server to hold GIS data for users. The architect connects desktop, server, and web applications to make one seamless impression of an otherwise complex underlying infrastructure.

Project managers, particularly in the consulting industry, are a great example of a career that can use GIS as a tool for project management. Project managers, as the name implies, are in charge of a project with a budget and need certain GIS tasks done for their project. In the example of a solar power plant, the project manager would be interested to know about, and be familiar with, data associated with their project. Some examples are, areas where solar panels can be placed, where wetlands are located, participating parcels and proximity to transmission lines. While they do not build or analyze data, they can work with and use the project data to make informed decisions and convey information to the client.

A field where GIS is used pretty heavily, is planning. **Planners** come out of university with GIS experience that can help them make better decisions for their community. They may want to look at traffic data for road upgrades or desolate urban areas for redevelopment. They may want to use census data to review planning around helping those most in need of community services. All of these, with regards to a city or county planner, are great applications and use of GIS.

The **military** uses a lot of GIS technology, although the information may only be partially revealed to the users. As was described by an ex-military GIS user, the work environment is extremely secure. Every action and operation was under surveillance. Operators used software in states of isolation, so that an entire mission objective would not be revealed. While the schooling was excellent and the applications can be challenging, the work sounded isolated and repetitive. One of the most fun things with GIS is knowing the overall objective and figuring out how to get there, rather than being fed pieces of the mission with the objective hidden.

Workers in the **green energy** sector, particularly wind and solar, have made extensive use of GIS technology. Most of our existing power plants and infrastructure have been in place for a while, but with green energy being a new wave of development, GIS is being used heavily to find suitable locations. In addition to finding new locations for a solar farm, a GIS is also a great tool for users in the solar industry to manage infrastructure. Knowing the status for equipment maintenance or other operational scheduling can all be queried and viewed in a GIS.

There are many more careers that can be listed. These users of GIS are interesting in that they have another primary job function, but use GIS technology to aid their efforts.

2.10 Geospatial Technologies: More than GIS

Geospatial has become a common term that encompasses GIS and the usage of the word started to emerge in the 21st century. Spatial is "relating to space" and geospatial is "relating to space on Earth". It's worth noting that the term geospatial has come to include GIS software and other related technologies. GIS is a large part of geospatial technologies. Today we see titles such as "Geospatial Analyst" that reflect this integration.

An example of a geospatial technology, that is definitely not GIS, is Global Positioning Systems (GPS). GPS is a network of satellites and ground locations that work together to provide accurate locations to GPS receivers. GPS is the technology we use to navigate on Google Maps or other location services on a smartphone. It has been simplified to a blue dot on our maps. Today, everyone has GPS in their smartphone and thus the capability to collect data that can be imported into a GIS.

Another example of a geospatial technology is LiDAR (Light Detection and Ranging). LiDAR is a remote sensing technology that is usually collected by an airplane or drone. The points are high density and can be used to create a very detailed Digital Elevation Model (DEM) in GIS. Other sensors in the aircraft can collect vegetation data that can provide information about chlorophyll, vegetation index classes, and near infrared to name a few. LASTools is an example of geospatial software designed to work with LiDAR point data and can also output various geospatial products.

Another example of open source geospatial software is the OGR/GDAL suite of tools. This software does a lot of different things, but it was so good at importing and exporting GIS data formats that even a large commercial GIS company had to ask the author if they could include the code in their software. GDAL/OGR are also used in open source GIS projects to import and export an enormous number of GIS file formats. Not only that but they have handy tools and functions for vector and raster data.[3]

To further promote the idea of a broader inclusion of technology, university GIST programs have gained popularity over the years, which speaks to the broader picture of geospatial technology. GIST is Geographic Information Systems and Technology; some name it Geospatial Information Systems and Technology. Notice the changing of Geographic to Geospatial and the addition of the word "technology". These are indications that GIS is a smaller subset of a broader technology umbrella.

---

[3]The Locate Press book, *Geospatial Power Tools*, highlights the power of the OGR/GDAL suite: https://locatepress.com/book/gpt

2.11 To GIS or not to GIS

Looking at outputs, often expressed as maps, GIS and Computer Aided Drafting (CAD) software may seem similar. They are similar in that they both work with points, lines, polygons and other spatial data in an x,y coordinate system. However, the intended use of a GIS and CAD software is quite different. Can you guess what the primary difference is?

If you guessed that it has to do with the "where" component, or spatial reference to the Earth, you guessed correctly. In addition to the locational component, GIS also utilizes advanced database technology to manage and query large datasets.

In this context we're talking about vanilla or plain CAD, not a spatially enabled CAD or one enabled with spatial functions. Computer Aided Drafting (CAD) is best for design work. CAD is used to build models of objects in computer space, to scale, such as a house or machine. When building computer models of real objects, there is no need for referencing locations on Earth. The model of whatever is being designed is independent of X and Y coordinates on Earth. This means that CAD can use its own coordinate system that is not referenced to the Earth but to the object being modeled. Without getting into the details of projections and coordinate systems, this simply means that data built in CAD cannot be used in GIS without changing the CAD coordinates into real-world coordinates.

A way to contrast the two software is the hypothetical situation of a Mars Rover deployment. If a rover or air drone is going to be sent to Mars, CAD is the perfect software to design the rover. Various configurations and prototypes can be viewed in CAD to engineer the rover itself. However, when the rover is deployed on Mars, spatial reference coordinates and terrain parameters become paramount. The design is done and now software is needed to analyze the terrain and manage the problem of how to get samples without getting stuck.

If the rover has to go from point A to point B, it has to know its location on Mars, points A and B, and also all the terrain variables in between point A and B. GIS software can be used to determine how to get to the destination with the least amount of trouble. Trouble in this case could be impassable areas, usually steep slopes or cliffs. Even this isn't the best example because it's not that much different from commercial online mapping software that guides one from point A to point B. It's still not a GIS, but the example illustrates design vs. location-aware technology.

Another misuse of GIS is the graphical component. The cartography aspect of

GIS involves a graphical component to make these maps. Making a map entails setting up a page and adding graphics and title blocks around the map. While there are graphical capabilities, using GIS to create graphics or non-spatial data presentations would be the wrong application for GIS. The obvious software for this task would be graphic illustrator software, another great piece of software but also lacking the locational component. Sometimes GIS and graphic illustration software are confused or conflated because a part of the process is making maps with pretty graphics.

2.12 Summary

Do we have a good feel for what GIS is now? It's software that works with "where" or locational data. It is applied to various sciences or disciplines and has two distinct user types: direct user or indirect user. Each user path determines if the person will become an expert with GIS or an expert in another profession that uses GIS software. The data, at a simple level, is points, lines, and polygons that have the locations stored as coordinates on Earth. These features have a database attached that describes the features with attributes and values. While this may not sound all that interesting, we find that many people introduced to GIS rarely say goodbye on their own accord.

3. What is Open Source & Proprietary Software?

3.1 Open Source Software

Free and Open Source Software (FOSS)

We should have a good idea about what GIS software is, and now comes the open source part. The short version says that Free and Open Source Software (FOSS) is not owned by anyone and is not written for profit. The software is free to use and the code is open to view, modify, and contribute back to. There are different open source software licenses that range from liberal to restrictive uses of the code. It's publically owned as opposed to proprietary, the opposite of what most users are exposed to. These are some key technical points of open source software, but let's try to explain it in a way that is less specific.

The term "open source" has evolved into a broader context of a philosophy or way to do something that is no cost, transparent, democratic and benefiting the greater good. Ignoring the legal and technical definitions, think of open source as just another way to say the process and code is transparent or open. The general public can look at how the software works, down to the exact code used. This would be in contrast to the proprietary nature of name-brand medicine, protected research and proprietary software which rarely, if ever, expose the details.

Generally, open source means the way the thing works is fully exposed through information; with open source software, it's exposed explicitly through code. Open source software development can be heavily rooted in academia with various high level projects used for masters and PhD work. This is really the cream of the crop as far as cutting edge applications and research. Sometimes these academic works of art become extensions or even part of the code base.

Some benefits of open source software are: no cost, based in academia—which provides best approaches and solutions, providing quick bug fixes and added features based on user needs. Most open source licenses work on some variation of the idea that anyone can use the code, but derivative work must be provided back to the community under the license.

> Because the software is open to use, the best way to experience the power of open source geospatial software is to get involved with it. Even if there is no thrill to be found at the command line where everything is simple and fast, the graphical interfaces are very user friendly and have an enormous amount of capability. While open source software was overlooked in the past, we now see mainstream adoption of desktop and web-based software.

Unmagical Software

Although, at times, FOSS does seemingly magical things, FOSS is not magical software; it is the result of people's hard work. An interesting note on open source software is something that was not realized until very late in the journey. Open source isn't a magical place where software is built in people's spare time, just to give us free stuff.

There is an interesting and yet somewhat hidden point of view expressed that the success of Linux, and other open source projects, came from a place of selfishness. Open source developers generally are not writing code for you; they are writing code to solve a problem that they have. They just happened to be so kind as to share these great software solutions with the rest of us. Why? Because then we get to build a community around it and build on the code.

All great things are accomplished with a team, with varying skills and experiences that evolve the code to higher place. The projects are unified through a public mailing list and code tracker that everyone can use to get and give information. Linus Torvalds, the creator of the Linux operating system, expresses this same opinion in an interview with BBC.[4] The same sentiments can be found on open source mailing lists if you pay close attention.

Further dispelling the magical spring of free software, there are sometimes crowdfunding efforts to compensate for the enormous amount of time developers put into the code.[5] If developers were just writing free code for you, there wouldn't be crowdfunding events to compensate them for their precious time. There is utmost respect for developers who make this entire open source existence possible by gluing us all together with code. Without them, there

---

[4]https://www.bbc.com/news/technology-18419231

would be no open source software to use.

Free or No Cost

Have you heard the saying, "The best things in life are free."? It's true. FOSS is free to download and use under any circumstances. It's not free because it has no value. People don't understand that having no cost is a feature of open source because the code is open and available to use. Free is a function of making the code publicly available. Why would it cost money to freely access the code needed to run the software?

Open source software is also free in the way of freedom, where users are free to do with the software as they please: view, modify, and contribute. For people rooted in the belief of capitalism, the idea of getting something valuable for free does not compute. This is certainly the thought amongst US business leaders, who do not understand what makes open source software special—and it's not the "free" price tag. They cannot comprehend that free software can have additional benefits, capabilities, and perform better than "equivalent" commercial-proprietary software costing over $14,000.

Throughout this journey and professional career, some amazing feats have been performed with open source GIS; some things commercial software could not do. In addition to capabilities, the performance of open source software is superb in comparison to clunky commercial software. It's hard to believe that there is free and extremely high quality GIS software that can do everything a commercial GIS can do, and better. The only way to dispel your own uncertainty is to download some and try for yourself.

Save How Much Money?!?

Why is commercial GIS software so expensive? It goes back to the UNIX days, when UNIX system administrators, the UNIX operating system and associated software was more expensive than a Windows "equivalent". Since GIS got its start on UNIX, so did the pricing. Funny, (not haha funny) how the prices just moved on over to Windows when the time came. Of course there wouldn't be a price adjustment downward, but looking at these costs paired with the mediocre software, there certainly should have been. It's really amazing that commercial GIS software is so expensive, when all it does is work with some points, lines, polygons, and attribute tables.

---

[5]Lutra Consulting runs a popular crowdfunding initiative for QGIS: https://www.lutraconsulting.co.uk/crowdfunding/

Small disclaimer, the price quotes are best recollection and go back two decades; so they may be off a little. The numbers are certainly close if not exact. Some real examples for GIS software savings over 20 years include a variety of geospatial software: desktop GIS, web mapping and GIS database.

On the desktop side for commercial software, one advanced GIS license cost $14,000 with $3,000 annual maintenance. Over twenty years, $60,000 for maintenance was paid, plus the initial $14,000, totaling $74,000 for one commercial GIS software license. If organizations had more than one license, the maintenance costs were about half for additional licenses— still not very affordable. Some organizations have dozens or hundreds of users. Add a few zeros to the price estimate for those organizations and this doesn't even take into account a business that scales globally. Did we mention open source software has no cost? None at all.

Extensions to commercial software was always a bit of a gimmick, at least from an open source perspective, because open source GIS software doesn't make you pay more to get more capabilities; it's just there. Core capabilities, such as working with raster data, are perfectly integrated into open source GIS software. However, commercial software charges more money to do more things. So, if you want to work with raster data, a license for a spatial analyst extension would be required. If memory serves correctly, it was about $2,500 with a $500 annual maintenance fee. In recent offerings, extensions are accessed through subscriptions. Uggh. Subscription-based extensions, how do they get away with this? Mostly because people don't know about open source alternatives. They don't even want to look because there are devoted fans of commercial software, similar to rooting for a favorite sports team. In this example, to calculate costs for extensions, spatial analyst, and 3D analyst were considered. Two extensions, totaling $5,000, with maintenance at $1,000 annually, totals $25,000 over 20 years.

Then, for web mapping, the purchase was $10,000 for one CPU, $20,000 if your windows machine had two CPUs. Why more per CPU? Is there any justification to charge more per CPU? Can't think of one; just seems like a cash grab. Non-technical people may see some justification for this, but from a technical perspective this doesn't make sense. On top of the initial purchase price, add in your maintenance because we can't forget that. Pretty sure that maintenance was $3,000, for Internet mapping capabilities, so half this number if that turns out to be the case. We're going with $3,000 over 20 years, another $60,000 for maintenance, totaling $70,000 over 20 years. A standalone Internet mapping software was deprecated as the commercial developer went to a hosted online system instead. This online system is web-based where users

are charged credits for usage and storage of data. The web mapping cost can't
be ignored, but can be adjusted in your own equation. Anyway we slice it,
commercial GIS software costs a whole lot of money.

Now for a GIS database. Sometimes this is called a spatial database and when
we implement one later on, it's referred to as a GIS data library. This uses Relational Database Management System (RDBMS) technology to provide enterprise scale solutions for otherwise problematic file-based GIS data. The open
source operating system Linux, Geoserver, the PostgreSQL RDBMS and the
PostGIS spatial data extension were all part of the cost savings. In just five
short years, there was a realized savings of $43,000 using an open source en-
terprise GIS spatial database.

So, over a 20 year period, for one user and some enterprise capabilities, there
is a savings of $74,000, $25,000, $70,000 and $43,000 respectively, a total of
$212,000. Sure, for the first year the commercial company includes main-
tenance (how generous) but we're using rough numbers here to illustrate a
point. The point is that over time, commercial GIS software fees erode prof-
itability. Initially, the cost of the software is enormous and the maintenance
seems like the lesser of the two evils, but over time it's clear that the maintenance far exceeds the cost of the software. What could organizations have
done with this money instead of paying maintenance fees? What could they
do going forward, given all the advancements in open source geospatial software? The answer is pretty much as unlimited as one's imagination.

In the early days of the journey, at a client site we were commiserating about
the annual fees for commercial software. It can be recalled that a large client
site with many licenses in the early 2000s, was boasting in a sort of forced
acknowledgement that they were paying over $100,000 a year in maintenance.
Some of this money was duplicate spending on the same type of extension
software— one for a UNIX workstation version and one for the more friendly
GUI for Windows. Open source GIS software clearly wins in the cost category,
sacrificing nothing in quality or capabilities.

Bug Fixes

In open source software, critical bug fixes occur faster than in a commercial
company and highly critical ones are dealt with as a priority. The community
files bugs, where they may be an end user or a developer. Tickets are filed in
an open ticket system and people can view the status of the repair and make
inquiries. Sounds great, right? It is.

Regular users can contribute to the code indirectly by filing tickets and testing
bug fixes. Sometimes end users will directly interact with the developers of

the software. Contrast these qualities with commercial software and it's clear which one should be better software in theory. The cool thing is, it is better in practice too. The more people that participate, the better the software becomes.

Extra Modules or Plugins

Plug-ins are an important way for software to stay current with technology and also fill special use applications the core software itself does not do. An example would be KMZ files that have an embedded HTML table in them. In QGIS, the KMZ is read without problem when querying the features, but in popular commercial software the embedded table is inaccessible. This means the attributes of the imported KMZ file cannot be queried. In QGIS, there is a KML (KMZ is just a compressed KML) tool plugin. It handles KMZ table imports and exports quite excellently, and hands it to you all wrapped in a pretty bow.

In open source, these plugins are based on user needs; no one writes a plugin they have no use for, unless it's for payment. Sometimes these plugins end up in the core software as was with r.stream.extract in GRASS GIS. This is an excellent way to expand the core software based on user needs and contributions, rather than bloating the software with features and functions no one cares about (eh hem, looking at you commercial software).

By contrast, commercial software can be slow to respond to user needs and requests in the form of plugins. Longer development cycles are less flexible and with bigger consequences to breaking the existing software. Plugins are an excellent way to work with modular code and test it by thousands of users to see if there are issues. Often with commercial software, user additions are not supported by the commercial GIS company, even if through a third party. There are user contributed scripts to the community, but not anywhere near as useful and well organized as in the open source world.

Community Support: The Best There Is

Commercial GIS software support was very good during the ARC/INFO workstation days (late 1990s), but eventually users were directed to online forums for help. These forums offload the resources of expensive technical support staff, which is provided with paid maintenance. In the forums, users solve each other's problems instead of their support staff. Sure, staff respond on the forums, but this cost saving measure removes the personal touch and timeliness of problem resolution. Knowledgeable support staff are expensive, but users helping other users is free. Originally, users called technical support and received helpful and direct answers. Now users search the web and post questions hoping for a response for their "paid" technical support through maintenance.

For current GIS users who are accustomed to helping other users through online forums, the transition to open source communication will be very easy. The community support comes from all around the globe, so there is a very wide user base to work with. With open source software, there is no 1-PAY-4SUPPRT number to call. There are commercial companies associated with open source projects that can offer user support, however, most support comes from the community.

The best place to find community support in open source projects is mailing lists. Mailing lists have been around since the early 1970s and are a great way to share information in a global network of users. Mailing lists might be a little frustrating at first, but it doesn't take long to get the hang of it. The responses can be very quick and most of the users are very well informed. Most open source communities use mailing lists to communicate about software project issues and topics.

In order for a user to get looped into a mailing list, they must subscribe to it. If they don't want to receive any more emails, they must unsubscribe to the list. When a new email thread is created, a user emails the list. The email then goes to all other users on the list. When users reply to the list, the reply goes to everyone. It is expected protocol to always reply to the entire mailing list rather than individuals. This ensures that information is kept 'on list' with the idea that information should be shared. It also ensures that the information can be retrieved for searching in the future . In all ways, the sharing of information is promoted in open source.

Another great thing about the support system in open source software is that the support can come from all over the globe and from different levels of expertise. Sometimes the support can come from the highest of levels, the developers themselves. This is entirely different from interacting with technical commercial support staff on a tier I or even tier II because they are not developers and they admittedly do not have access to them. Developers know how the code works and can answer specific questions about the software. The fact that they are so integral to open source projects and can be accessible to users is truly amazing.

It's rare, but sometimes there is no answer to a question asked on a mailing list. Maybe something has been misunderstood in the post or the one or two people who have encountered the issue are not available. However, experience has shown that someone, usually a steward of the software, sweeps the lists looking for unanswered questions to make sure everyone has the help they need. It's not a given, but it has been observed. It's important to remember there really are no silly questions and the people involved in the community enjoy welcoming new people and helping each other out. The number one rule is to be respectful.

Some Open Source GIS Software Names

Now we add GIS to this open source software discussion. By extension of open source to GIS, it seems that open source GIS software is just free and transparent software for GIS users. In all reality, unless one is a developer, having access to the code isn't all that useful to most users. Some of the primary benefits of open source software don't seem to resonate with average users, but that's okay because the benefits can be indirect.

The value of open source GIS software is not only the quality software, but there is much value in becoming part of a community and absorbing the philosophy of open source software. It's contagious, as new users get familiar with the community it isn't long before they are helping other people in the GIS community. No information is protected or hidden in the open source world and this creates a great environment to learn and share ideas about how to do things. It also ensures that best practices are performed and optimal results are obtained.

It's great to talk about open source software in general, but we want names. One open source desktop GIS stands out with extreme popularity: **QGIS**. It was originally known as Quantum GIS and later shortened to QGIS. This software has come a long way since the early 2000s.[6] There is an incredible community of users and developers backing this project. The plugins, contributed by users, are an amazing testimony to the creativity and global involvement of QGIS users. QGIS has been around for over 20 years, but over the last 10 years this software has really matured and won the affection of many users and developers. QGIS is free to download and use for personal or commercial use. The only real request is that you get involved with the community in some way. QGIS is really fantastic; it has a slick interface, a massive selection of user contributed plugins and tight integration with GRASS GIS.

GRASS GIS is a powerful GIS that has roots (no pun intended, okay it was) back to the 1990s. This software is very stable and munches on big data like a hungry kid at the movie theater with a giant bucket of popcorn. The design is very modular and very command line friendly, following the UNIX philosophy. The UNIX philosophy is an idea to write simple small programs that do a

---

[6]QGIS was founded by Locate Press author and former publisher, Gary Sherman, as a way to view map data from PostGIS in a desktop app.

thing well, later to be combined with other small programs performing larger tasks. It is easy to write scripts in GRASS GIS and you can pick your own language. The selection of scripting and programming languages is wide, with Bash, Perl and Python to name a few.

There are also open source utility programs like **GDAL/OGR** and **LASTools**. GDAL is a suite of tools for vector and raster data. It is used by QGIS for many functions and GRASS GIS too. LASTools is an amazing command line tool for LiDAR data. It provides tools that can extract LiDAR classes and reproject data. These open source geospatial tools plug right into open source software and immediately extend capabilities.

> The above tools provide a powerful platform for analyzing data, running analysis, and creating various outputs (including maps). Other tools such as web mapping products, spatial databases, cataloging systems, and more, are available and some are discussed later on in the context of the solutions they provided.

Open Source Software Challenges

Open source software is really fantastic, however, it can present some challenges to new users. Technical knowledge can help overcome some of the struggles, but other issues like awareness require more time for saturation. During the open source GIS journey, several of these challenges were encountered and resolved or determined to be a non-issue.

Awareness

One of the biggest challenges to open source (GIS) software is awareness. "What is it?", and "How can good software be free?" are two common beginner questions. Commercial GIS companies have marketing teams and clever business interactions to hype their software, while open source software has passionate users and highly skilled developers (also passionate developers and highly skilled users). Proposing an open source solution can be met with great skepticism and even rejection if an organization has already purchased large amounts of commercial GIS software. Awareness has been increasing, by way of user groups, social media, FOSS4G annual user conferences, and companies dedicated to open source software books, such as Locate Press.

Bugs

The bugs in open source software aren't really a challenge, users just need to know there is a different way to deal with them. Resolving bugs in open source software requires users to take some action by asking the mailing list or filing a bug report in the tracker. This is different from users paying for commercial software with expectations that the company will resolve issues. Open source software has an advantage where users can place bounties on development requests. The large amounts of money saved on commercial GIS purchases can be used to develop the software and obtain custom solutions.

Technical Support

The challenge with techncial support has been met with social media and sites like Stack Exchange, where users can post questions and get answers from the community.[7] There is no direct technical support to call in open source GIS, but there are open source GIS consulting companies who are available to fill this void. Commercial GIS companies have been tapping into the web as a resource, where users help other users or a commercial GIS employee posts a repsonse in an online forum. By doing this, the benefit of commercial GIS support has become less visible since mailing lists and online forums can help open source users too.

Preference of Technical Knowledge

The people who write code are very smart and it's been observed that technical expertise can be taken for granted. For example, a user may feel totally lost when they ask the list about a problem and the answer is, "You need to download the code and compile the GIS software against the latest library.". While it can be rewarding to get this stuff working, it can also be frustrating resolving software issues that are taking time away from GIS tasks. Don't forget there are some really good GIS consulting companies that can help in these areas too. The last thing anyone should be in open source, is alone.

Suggestions to Help

A good way to avoid these challenges and have the least amount of issues with open source GIS software is to work with commonly used software. Open source GIS software links with other open source software and libraries. When other linked software is mismatched, known as broken dependancies, the GIS

---

[7]https://stackoverflow.com/questions/tagged/gis

software may not work. Repositories store software that is known to link and work with other software in the repository. Working with commonly used repositories is a great way to access the most well tested and debugged code.

Along these lines, the top recommendation is that once you have open source software working on Linux, leave it alone. Don't be tempted to upgrade unless you: 1) are willing to take the risk of broken software, or 2) must upgrade to fix a critical bug (and still #1). Compiling code and linking software on linux can be complex and one curious misstep and the next thing you know, a day has been spent trying to get code to compile or locate a missing library.

Some people just can't wait to try that new tool or enhancement and in exchange for this lack of patience they may get to spend a lot of time resolving dependencies. Being one of these impatient Linux users, it was fun installing software and getting it to work. One thing is for sure: a lot was learned in the process. Staying with mainstream software helps ensure that all the different GIS software packages will work together without issues because they have the most amount of usage, exposure and thus fixes.[8]

3.2 Commercial & Proprietary Software

What is it? The Antithesis of FOSS

Most people are familiar with commercial software and it is worth discussing it to show how much it contrasts with open source software. Commercial and proprietary software is the exact opposite of open source software. It's unfortunately all that most people have been exposed to. Commercial software is written for profit and the code is private. Rarely are commercial and proprietary software mutually exclusive, except where companies may produce open source software as part of a broader type of service.

How many random operating system reboots or crashes have users had with the world's most popular desktop operating system? It's easy to be frustrated with this erratic computer behavior but most users overlook it and consider it acceptable because it's assumed "it's just the way it is" or "what else could I use anyhow?". In our context, as we journey forward, we're talking about the dominant elephant in the room for commercial GIS software, which shall not be named.

For some reason, the vast majority of commercial GIS software users have

---

[8]OSGeoLive is a convenient package of all of the most popular open source GIS applications: https://live.osgeo.org/en/index.html

come to overlook the bugs, performance issues, and mediocre to terrible technical support. In many cases, they are forced to use commercial software and be unhappy. In this journey, sentiments have been developed through direct interaction with both commercial GIS software and open source software, and a supportive community full of enthusiasm. As the commercial GIS software grew in popularity, it became clear that there was an overwhelming number of commercial GIS users who realized the same thing: commercial GIS software stinks.

Benefits of Commercial Software

The main benefit to using commercial GIS software is that it has widespread use. This means good word matches with job descriptions. Being taught the commercial software brand in university was beneficial in being able to directly meet requirements for job opportunities. A secondary benefit to using commercial GIS software is that the commercial company glues the environment together with other software they sell for web mapping and cloud hosting. While this sounds helpful, the perils of vendor lock-in are discussed later.

Also, because of its dominance, solutions for commercial GIS software can be readily found on the Internet. With paid annual fees, users also have access to technical support staff that can help with software questions or problems. Commercial software has a goal to write software in a way that anyone can start working with the data and start creating stuff. It's that expectation of ready-to-go software when paying a hefty price.

There were frequent discussions on mailing lists from students and professors, talking about which software to teach in college or university. When students were taught commercial software, there was a relatively easy match for job requirements and skills. On the other hand, there was this liberating software that could be used for free, with cutting edge technology, and an amazing community of support. They both express the concepts of GIS; which one should be taught in a course?

It seemed as if teaching commercial software was doing the most benefit to students by placing them with jobs, but the bigger picture about vendor lock-in and expensive software was ignored. Open source software had a better philosophy, community and software. Having gone through extensive coursework with commercial GIS, and being saturated in both types of software, the opinion was formed that teaching both software would be beneficial. For commercial software, it would provide the experience matching market needs. Open source software would empower the student to offer their employer huge cost savings and a more open data system for their organization.

Why Commercial Software Loses Against FOSS

In the early 2000s, a new commercial desktop GIS was released for Windows, the bugs were so problematic that it had to be put back on the shelf for production use. Software that is commercial and proprietary is often slower to respond to user and market needs. An opposing philosophy is open source software, where the source code is made available to everyone for rapid development. This allows many people to fix bugs and introduce new features.

Closed software is generally slower to fix bugs and slower to adopt new and necessary features into its software offerings. A constant response from commercial GIS software representatives was, "We are aware of the bug, we are aware it has persisted through several versions, but you will have to wait until the next release for the fix." Waiting for software fixes became among the most frustrating aspects of working with commercial GIS software. The bugs were frequent and users were constantly trying to figure out problems with the software. As imagined, this wasted an enormous amount of time.

Commercial software can be slow to respond to user needs and requests in the form of plugins or add-ons. Longer development cycles have bigger consequences to users who must wait for fixes. Plugins are an excellent way to expand a software's capabilities and adapt to new formats or create specialized functions. User contributed plugins aren't as common in commercial GIS software, because they want you to buy their plugin or extension, not get one from the community.

Proprietary data formats are another area that lose to open data formats. This is discussed in detail in a later chapter via an Enterprise GIS Data Library. GIS is not a specific software vendor. When a commercial software vendor makes proprietary data formats, it is with the devious intent to lock people out of data unless they have the special key. The special key in this instance is their expensive GIS software. Let that soak in for a moment, in a technology based in science, commercial GIS companies are looking to profit by using proprietary data formats. Disgusting.

GIS is nothing without its data and when the data is locked behind a data format, it cannot be easily shared or used by everyone. Thus, closed data formats hurt the industry because GIS data isn't meant to be used by one person. GIS data needs to be open for everyone to access, so it can be used and validated by as many people and organizations as possible. This strengthens the data and helps us all do better work for whatever the purpose. Public data can be used for a lot of good to help the environment and local community.

Similarly, companies that support open source can also do a lot of good. It's worth clarifying that the actual software code can be open source or proprietary (closed), but business applications can be both commercial and open source. For example, it's not uncommon to find companies that offer consulting or technical support services for open source software—the best of both worlds. In many of the discussions with the big commercial GIS company, a constant utterance was, "But you can't make any money with open source software." It's certainly true that it can be difficult, if not impossible, to charge money for software code that can be readily downloaded. However, there is evidence that proves commercial entities can profit by providing services around open source software.

Cornering the Market: Supply and Demand

So why is commercial software used? In the case of the commercial GIS software monopoly, it's a well designed feedback loop. The simple answer is that commercial companies spend far more money on marketing when they should have spent the money on developing a better software product. Digging deeper, we find that specialized marketing tactics allow commercial software to be embedded in large organizations, government, and educational institutions.

The first advantage gained by making commercial GIS software sales to a large organization is the sale itself. There are instantly new users, sales of software and annual maintenance to consider. Providing a discount or even giving away software to educational institutions may seem quite generous, but the annual maintenance will soon be scooping money into the commercial GIS bank account. This discounted software has far greater payoffs than the initial purchase savings.

The second advantage gained by a commercial GIS software sale may not be so apparent. The large organizations that buy the GIS software then force other people and other organizations to buy the software. How? Without consideration or hesitation, these large organizations will ask for data to be delivered in the commercial data format they have bought into. If another organization wants to do business with the larger organization that has the commercial software, they must buy it too. Only the commercial GIS software can create the proprietary commercial GIS data formats and GIS project files. Large organizations go down this road because: 1) they naturally follow their commercial software investment, and 2) some don't know any better. But is ignorance an excuse if the organization is a large public entity?

For example, a state agency may put out a bid requiring that deliverables be in a specific GIS commercial software format. This sets up the market to be biased towards the commercial GIS company. Corporations may be more independent in the way of not needing to partner with someone to do the work, but for some reason corporations are easily duped by other corporations into thinking they need some corporate GIS solution. Either way, the demand has been created: GIS jobs using commercial GIS software. Now to rig the supply side, the students.

The supply side is the educational institutions. Sometimes large software discounts or free pricing are provided under the guise of helping education. The obvious benefit is that students are taught a specific commercial software for what is otherwise general GIS concepts. For example, in the mid 1990s there was a book used to teach in university called "The ARC/INFO Method".[9] ARC/INFO is old commercial software that was deprecated around the year 2000. This book was used to teach GIS and required the use of specific commercial commands and syntax.

Students should be focused on creativity, learning and understanding broad applications. Instead, students are learning a specific commercial software that prepares them for exact job matches in the marketplace. This sounds wonderful for everyone, so what's the problem? The problem is all the pitfalls that go along with commercial software. Students typically have no idea about this commercial entrapment, as they are a blank slate for learning and likely new to the field. The problem is that other people have made decisions that shape their future and limit their thinking, instead of expanding it—hardly the hallmark of a liberal university education!

This is how the feedback loop of commercial software is built: Educate new users in educational institutions with commercial software, then ensure the market is primed with the same commercial software demand. Voila! The perfect feedback loop to create an eternal customer cash flow. Trap organizations with vendor lock-in and there will be indefinite cash payments. The system traps users and organizations with commercial software and the annual recurring fees. We believe this behavior is monopolistic in nature and guarantees users end up using the most well-known name brand, not the highest quality and most useful software available.

So Why Use This Rubbish?

People use this rubbish because they're trapped. This all sounds relatively harmless until we realize that these large organizations, commercial, government and educational, are all strategically solicited by commercial GIS companies. Some key members of institutions were taken out to lunch, while others were given reduced rates. Working at a small firm, we were never asked out

---

[9]https://www.amazon.com/dp/1879102005/

to lunch, invited to special occasions, given any special provisions or offered reduced costs for our firm. It was actually quite the opposite; we were often laughed at and mocked when issues with the software were repeatedly brought up to production supervisors and regional representatives. Their reply, "Well, what other GIS software do you think you're going to use?"

Why monopolies are bad for consumers is another book entirely, but there's a parallel between the association people have with commercial operating system software and computers in general. When people say they have a computer, the assumption is that it is using Windows. Granted this is changing but the same analogy is for GIS software. When people say they use GIS, the assumption is that it is the dominant commercial GIS software. At least this is true in the US, hopefully not elsewhere. These monopolies overwhelm users and market them with sparkling objects. If everyone assumes GIS software is only one commercial GIS software offering, then they would never think or know to look elsewhere. In fact, the propaganda is so strong that some people laugh or scoff at open source software just because it is free. Turns out, the joke is on them.

Trapped by Data

Commercial software companies further add a moat around their castle with proprietary data formats. This locks organizations into their software and locks out other users from accessing the data in an open format. Proprietary data should in no way be considered a format acceptable to distribute public data to users. Geographic data should be in a format that all GIS can import and work with, rather than requiring the public to purchase software that unlocks the data.

Closed data formats have a long history with the commercial GIS software company, and eventually people figure out how to read data from the formats; so why try to keep this data a big secret? It was interesting to watch commercial GIS try to take over an open data format for LiDAR, known as .las files. To some, creating proprietary data formats is a crime against the GIS profession. Why? Because it discourages the sharing and exchanging of GIS data so that profits can be made.

Take, for example, the historical war on LiDAR data. When LiDAR data exploded in use, the commercial software company tried to make a proprietary LiDAR format. They also made some very interesting claims about performance. In short, they lost the LAZ Clone Wars. The longer story can be read, by honorably mentioning the author of some really nice LiDAR tools, Martin

Isenberg.[10]

3.3   Do We Have an Answer?

We made it to the end of the chapter, but has the question, "What is open source GIS software?" been answered? On a simple level, GIS is just software, an applied technology to the sciences. When digging deeper we see that it is special software designed to work with spatial data. At a basic level, spatial data is points, lines, or polygons with X and Y coordinates relating to Earth. GIS software must be able to work with the geometry and attributes in a GIS layer.

As far as the type of software, there is commercial and open source GIS software. Commercial software is expensive, closed, inflexible, and slow to make changes and fix bugs. Open source software is a thriving community of people, both users and developers, who love to work with GIS software without commercial entanglements. Open source GIS software is high quality, user involved GIS software that is extremely adaptive and flexible to user needs. The development cycles are solid and consistent, with rapid bug fixes and a community built around helping each other.

That's it for the overview of GIS and open source software. With these two topics behind us, we're ready to get into the good stuff. In the next chapter we learn about GIS technology and data as we go through GIS 101.

---

[10]Keeping   ESRI   Honest: `https://rapidlasso.com/2014/11/06/`
`keeping-esri-honest/` and LAZ Clone Wars: `https://rapidlasso.com/2014/02/`
`01/clone-wars-and-drone-fights/.`

4. GIS 101: The Academic Basics

4.1 The Journey Begins

Not much of a journey so far, huh? It will be helpful if we have the same basic information going forward about GIS and open source software. Beyond definitions, we hoped to establish a basic understanding about what a GIS is and some examples showing what it can do and who uses it. This chapter is a *GIS 101* course, providing the basics taught in university or an instructional class.

Of course university is an amazing place to study GIS, but they probably won't teach about how to cope with the insecure IT guy or how unfortunate situations in life can turn out to be the best in retrospect. They might also be ill equipped or uninformed about open source GIS software too. The stories embedded within this journey are intended to wrap up technical information and concepts into a non-technical book.

4.2 Various Disciplines with Various Applications

GIS users start in different places. Not only do these users progress to basic and advanced levels of GIS usage, but each person's background is unique in shaping how they use and view GIS software. The GIS field is a hodgepodge of various users unified by geospatial software. GIS software is an applied technology to the sciences and this means that the variation in users, and how users apply GIS, can be quite wild.

Users with a technical background will migrate into different areas than users with a non-technical background. A technical background in this context refers to topics like: computer programming, databases, networking, security, hardware, software, servers, web technology and mobile applications. Having a technical background drastically shapes one's journey with GIS software because GIS draws on many technological components beyond spatial mapping.

Every user has their own unique GIS journey based on the profession they choose and their background knowledge. For some, the GIS journey starts in

college, while others have already begun their career and then encountered
GIS software. Some will use the software as a primary expertise while others
will use the software to help them with their area of expertise. No matter the
variables relating to how we get involved and use GIS software, we all share
and use the same basic GIS 101 knowledge.

4.3   Preparing for the Journey

As a precursor to the journey, a lot of time was spent with computers in the
1980s. One of the fun projects with Dad was soldering computer chips on a
motherboard to build an Apple IIe computer. A new fascination was immediately found when the project was complete: the computer. We used dial-
up modems to connect with other computers and communicate on electronic
bulletin boards. There wasn't a usable internet for the public yet; there's no
recollection of web browsing on the Apple IIe. We played old school video
games like space invaders and a text-based adventure game.

It was an unforgettable moment the day Dad came home with a box the size
of a toaster oven. "What's that?" a curious kid asked. "A 10MB hard disk
drive" Dad said with a big smile. He added, "We don't have to use those tape
machines anymore." Playing with computers grew into something so much
more. It can't be stressed enough how a few childhood experiences can shape
a kid's journey for a lifetime.

Interest in computers continued to grow. BASIC and PASCAL programming
languages were taken in high school and they were a lot of fun. The teacher
provided a letter of recommendation to college admissions, with accolades
about the work done in the classes. This experience with programming proved
to be very useful in the future with GIS software. It's important for new users
to identify that too; what they like, what they are passionate about and what
drives their curiosity. GIS is such an enormous field that there's plenty of room
for all kinds of users, with all kinds of backgrounds and interests.

4.4   A Simple Suggestion Started It All

In university, during the mid 1990s, GIS was mostly a hidden topic. Full dis-
closure and interesting note, Stockton College of New Jersey transitioned to a
University and so did the degree. Nice upgrade. When this personal journey
began in 1995, the university taught GIS as a subsection or focus area under
the Environmental Studies degree. Geography is another common degree that
has a focus area in GIS. At that time, there were schools and professors teaching GIS, but it had not reached the popularity it has today.

It was now 1995 and it was junior year at Stockton University. By this time, it had been determined to pursue an environmental studies degree. It wasn't known at the time, but looking back, Stockton was a progressive university. Stockton had a functioning geothermal well field and these mysterious GIS courses. The school was very fortunate to have an an excellent professor teaching the GIS concepts and applications.

As an Environmental Studies degree was nearly in reach, another student had mentioned that the Introduction to GIS was a great class for environmental students. GIS? What is GIS? The course description was a bit cryptic, talking about spatial analysis and cartography. It sounded like it had something to do with computers and an application to environmental sciences. Both of which were interesting, so let's go.

Without much understanding about what was going to occur, the intro class was taken. It was called "Intro" to GIS, but a lot of material was covered for an intro class. It wasn't cartography and how to make maps, it was in-depth GIS studies. The courses taught GIS through ARC/INFO on UNIX workstations. This is an interesting origin story to note because, spoiler alert, the journey starts with UNIX and goes full circle to an open source implementation of UNIX called Linux.

4.5   GIS College Coursework

In total, there were three courses available to take at the university: *Cartography*, *Introduction to GIS*, and *Advanced GIS*. The Introduction to GIS lecture was both overwhelming and interesting, and the lab proved to be thorough, challenging, and a lot of fun. The hands-on lab really made GIS come alive. The labs reinforced the lectures, allowing students to play with points, lines, and polygons. The next course was Advanced GIS with a lab, which dove into raster data and real-world GIS applications. Stockton University had an atypical four credit lecture with a one credit lab, which provided a lot of hours and depth to the course work. After intro and advanced courses were completed, the last course, Cartography, was taken in the final semester.

The specific book chosen to teach both Intro and Advanced GIS at the university was, "GIS: The ARC/INFO Method".[11] ARC/INFO was a very expensive commercial GIS software, and in all fairness, the book was really good at illustrating GIS concepts and applications. The book taught all of the important concepts and the corresponding hands-on labs were excellent at illustrating GIS concepts. Unbeknownst to the book, it unfortuntately guided the GIS market into a closed loop of cash-flowing customers. It's not the book's fault, it did a good job educating students. In retrospect, the ARC/INFO "method" is silly because all the ARC/INFO "methods" were really just general GIS concepts and tools: buffers, clips, vector and raster data.

Intro to GIS laid the foundation for this GIS journey, providing a solid knowledge base from textbooks, classroom instruction, and a heavily weighted hands-on lab. In the mid 1990s, the course was well put together on UNIX and it seemed reasonable that a university might choose a leading commercial software to teach, due to marketshare and probability of being there in the future. So while possible, in the 1990s it might not have been the best choice to use other open source GIS software given the relative newness to the courses being taught. As open source GIS software matured, the debate about which one to teach, commercial vs. open source, became a very good question.

As the courses were taught, they were taught using commands specific to the commercial GIS. This turned out to be just fine later on when open source GIS software was encountered, because the theory was the same, there were just some new commands to learn. By only using commercial GIS software it tended to condition students to believe that GIS was always an expensive piece of commercial software.

Due to an excellent uptake of GIS concepts and working with UNIX, the GIS professor made an offer to be the GIS lab assistant. This was the first time teaching students in a formal capacity and it was discovered that teaching was really enjoyable. Something that would always be remembered and revisited when tutoring a PhD candidate in GIS technology many years later.

4.6   GIS Vector Data Essentials

GIS Data Types

Much of the Intro to GIS course taught what a GIS is, a system to store, manipulate, and analyze geographic data, blah, blah, blah; we already covered this in a previous chapter. Once we dive a little deeper, the next step is to get into some data. GIS software is interesting with all it can do, but the data is a critical component sometimes needing to be manipulated or converted into another type for GIS analysis.

---

[1]https://www.amazon.com/dp/1879102005/

The two most common types of GIS data are vector and raster data. There are also 3D voxels, short for volumetric pixels. Side note on voxels, it's funny (the laughable haha funny) because the commercial GIS company used their own 3D modeling structure for decades, but at a recent user conference the president announced this "new" model called a voxel. Turns out that voxels have been in open source GIS software for well over a decade; maybe two. We won't get much into voxels other than maybe an example modeling an underground contamination plume or making maps in Minecraft.

We'll start with vector data because it's the easiest to understand and work with. Vector data in GIS is exactly the same as vector data studied in geometry class. Vector data is discrete, which means it has a specific location on Earth or has definitive boundaries. A point is an X,Y location, a line is a collection of at least two points and a polygon is a closed line defining a boundary or area with at least three lines; or a curve.

Points, lines, and polygon shapes represent real world features in a GIS. Quite conveniently, these shapes can map everything this Earth has on it and it's up to the user to use the correct geometry type. It is usually not difficult to determine when to use a point, line, or polygon to represent a feature. Some examples are: points to map well locations, lines to map streams, and polygons to map city boundaries.

It should be a rather straightforward process to select the correct point, line, or polygon feature for mapping, however, sometimes there are situations that require additional considerations to arrive at the best feature representation. Suppose we had to setback 50 feet from a road edge. A line can be used to map a road as a centerline and we can buffer 50 feet from this centerline. However, a road is really an area, not a line as it is typically illustrated.

To be more specific about the extent or limits of a road, we would typically map the limits of the road by the limits of the right-of-way. The right-of-way is the width of the drivable road, plus the shoulder and wider. The area beyond the paved road is typically where utilities are placed or reserved for widening the road in the future. The right-of-way varies per road type, but on average we can estimate them to be about 66 feet wide. If we had to setback from the road 50 feet, using a centerline would not account for the approximate 33 feet of right-of-way on one side. The setback distance would fall short of the actual value by 33 feet.

A similar situation applies to streams. Streams are typically illustrated as lines, but streams have an area like a road. A stream is defined as having banks and the stream is actually contained between these banks. For most purposes, representing a stream as a line feature is just fine, but some regulations may require defining the stream as a polygon between the two banks. Just pick what seems reasonable at the time. If you find that you are having trouble, or bending the data to get the desired outcome, try a different way to represent the feature.

A Word About Topology

Ah, right into the abstract next level stuff, topology. We hit this topic up front because the feature definitions will reference it. In university we were taught, "topology is the relationship of features to other features". Memorizing the definition was easy, but understanding what it meant was a little more difficult. When we talk about data topology, we are referring to how features interact with each other *within a GIS layer*. A noted exception, topology rules can be extended to include relationships to other layers (e.g., must overlap), but topology mostly relates to the spatial integrity of features within the same GIS layer.

For example, adjacent polygons share the same boundary line and the software is aware of this. Topology is primarily a concern, or should be, while editing GIS vector data. It's not a concern when viewing the data or even using the data for analytical overlays, other than if the data is topologically incorrect, it may not view or overlay as expected.

Points

Each of the vector data models build from the simpler geometry type before it. Let's start with the simplest element of vector data, the point. A point consists of an x and y coordinate for its location. That's it. It's just like they taught it in geometry class. If we go back to geometry class, we can use a grid starting at a 0,0 origin. If a point is located at 5,7 it means that it is five units to the right (east) of the origin (our X coordinate) and seven units up (north) from the origin (Y). Hopefully it's obvious that point geometry does not have a length or area associated with it. Pretty simple so far, right?

Points are simple features and thus there aren't many situations for data violations or topology errors. There is one constraint, which is that points should not overlap. Overlapping points can be acceptable if they have differing Z values. An example would be sample points in a well. The points have the same X and Y locations but the Z value, or sample elevation value, differs. Data in this format, same X and Y but different Z value, is useful for delineating a 3D subsurface contaminant.

Lines

In GIS, lines are technically line segments. Line segments have endpoints, while lines extend straight in both directions infinitely. In GIS, the two endpoints of a line are called nodes. A vertex is a point along the line that defines the shape of the line. Lines have a length associated with them, but no area. Got it?

There are different types of nodes, some of which can be errors in a GIS layer. Note that there may be slight variations among GIS software in the exact terms. A true node is a point where three line endpoints meet. True nodes have no errors associated with them. The software can take care of the node locations being exact by way of *snapping*.

A *dangling node* is the end of a line that doesn't connect to another feature. In a line layer representing roads, we may have dangling nodes. Are they invalid? It depends. A dead end would be a valid dangling node in a line layer. However, if we consider a highway, there should be no dangling nodes along the highway because it's one connected road line with no breaks. Jumping ahead, a polygon is made up of closed lines, so an unconnected line in a polygon layer is an invalid dangling node.

A *pseudo node* is a node that touches one other node. So in this instance, we can use our road example again. Why would there be a node between two lines in a road network? Shouldn't that just be a vertex? Maybe it should be, but there are a few reasons that a node might be necessary.

One reason the line may need a node is to signal the beginning or end of the line. In our example, we can keep it simple. If the road changes names, or has alternate names or attributes, a new line feature needs to represent this. There could be a Route 9 which is also known as "Main Street" in the town. For a line to have different attributes, it needs to be a separate line.

Differing attributes are one reason a pseudo node may exist in data. Another reason may be a logical break in the data which could be as superficial as needing to label shorter segments. In the days of ARC/INFO, there was another reason to have valid pseudo nodes; the software had a limit on the number of vertices a line could have.

There are general rules for lines that every GIS software follows. One rule, though meant to be broken, relates to intersecting lines. In general, lines that cross each other should be broken at the intersection with a node. Digitizing a bunch of lines that cross each other and are not broken at the intersections is known as spaghetti digitizing.

Spaghetti digitizing is a no-no, but there are use cases where lines should not be broken on every intersection. One example is a road network used for routing cars from point A to point B. Entry and exit ramps may loop over or under the highway, because in reality drivers just can't turn onto a highway. When routing a car from point A to point B, this kind of behavior can be applied to the routing model because it forces cars to take the longer ramp loop rather than just route at a broken node that touches the highway.

Polygons

A polygon is an area defined by a closed line. This closed line is composed of points, which we know as nodes and vertices. A polygon must have at least two nodes and one vertex, and the polygon edges, boundaries, must not self–intersect. Adjacent polygons should not overlap or have a gap unless intentional. There's one other rule: the start and end nodes must be on the same X,Y coordinate. This ensures that the shape is closed. A polygon has an area and it also has a length as calculated by its boundary. A common GIS polygon feature is a property boundary. Other examples can be land use or wetland polygons.

Errors in polygon layers come in many forms, but the two most common errors are: overlapping polygons and gaps between polygons. Overlapping polygon features are exactly that, adjacent polygons that don't share the same boundary. There are a lot of interesting scenarios that arise while mapping features in a GIS. If a situation arises where the user feels there must be overlapping polygons, make a new GIS layer instead.

While there is never a case for overlapping polygons (yes, never), there is a case for gaps in a polygon layer. Gaps are areas between polygon boundaries. Any polygon without an adjacent polygon has a gap by definition. So, a single polygon parcel boundary has a gap topology error and it should be ignored or marked as an exception. Another type of gap is an *island*, similar to a donut hole. The donut hole is empty, meaning it has no polygon there. Two gap topology errors exist in a single donut polygon, the outer donut boundary, and the inner donut hole boundary.

More Than a Word About Topology

In university, *topology* was taught as best practice for editing vector data because it ensures data integrity. Actually, it was taught as the only practice for editing vector data. Is it still? Doesn't seem like it based on experience with new GIS users. Topology is an advanced topic that requires the right software and a data format that supports topology.

Today, most commercial GIS users don't have the expensive software they need to use topology features. One of the most commonly used GIS files in commercial GIS software is the shapefile. This is one of those commercial GIS company data formats that lacks topology. As a result, most users make use of hack-and-merge techniques, where polygons are created and then cut to prevent overlaps and gaps in polygon data.

Overlaps and gaps may seem unimportant, but they can present a problem when summary tables are generated from topologically bad data. Suppose you own a piece of property that is exactly 10.0 acres or 43,560 square feet. Let's suppose further that the property has 5.0 acres of pasture and 5.0 acres of wetlands. The total acres of the pasture and wetlands should be 10.0. This seems obvious, but if the polygons overlap the total could be 10.1 acres and if they have a gap the summary could be 9.9 acres.

"What's the big deal, we'll just round it in the spreadsheet?", said the young, energetic and naive GIS user. Altering exported data and not the source GIS data is a really bad idea. Further queries to the GIS data will give different results than the published data table. For small project areas, summarizing gaps and overlapping polygons may not make a difference in table summaries. Large projects with many cumulative overlaps or gaps may have enough issues to cause significant discrepancies.

GRASS GIS is an open source GIS that uses topology. When editing the data, the symbols change from a standard polygon with an outline and fill, to a line (called a boundary) and a point. More specifically, a boundary and a centroid — not a true centroid — these centroids are forced to be inside the polygon. The software knows that the boundary and centroid make up the polygon and that the attributes are associated with the centroid. When in edit mode, the software highlights incorrect boundaries and centroid topology, but when in read mode, these features look like a regular polygon from geometry. Users must ensure that the boundaries are closed and a centroid is within each feature for proper polygon topology.

Let's suppose there are two square polygons adjacent to each other with a shared common boundary. If the data format does not support topology, such as a shapefile, then the two adjacent polygons will have a duplicate line at common boundaries. The attributes will be stored in a polygon feature.

For data and software that support topology, there will be one line to store the boundary for the two polygons. The software will build topology and maintain an internal table that identifies which boundaries and centroids are part of which polygons. It will also track which side of the line the polygon is on. Lines have directions associated with them, so a left and right side of the line can be determined. In edit mode, the boundaries are shared between polygons, but there is no polygon, so a centroid is used to store attributes. There is no polygon until the user exits edit mode and becomes read only.

There are some odd cases such as the island polygon or donut hole. In the donut and donut hole example, the donut hole is not a feature. When the donut hole is clicked on, there is no feature there. When in edit mode, the software will identify the donut hole as a topology error because it is a boundary without a centroid. There is no problem, and the user has to know this as part of working with the data.

GIS Database Attributes

GIS data has two major components: the geometry discussed above and a database table. Feature attributes are stored in a connected database as fields or columns. The database aspect of GIS may be more familiar to people, as most people have encountered a spreadsheet table before. Attributes in the table describe features in a GIS layer. For example, a polygon land cover layer may have an attribute type which holds text values of: *wetlands, forests, pastures,* and *shrubs.*

The GIS software manages the database aspect of the data. The database used by GIS can be simple to complex. If you haven't worked with databases, maybe you've worked with spreadsheets; it's the same idea except SQL is used. A table can be opened for each layer and the feature records can be reviewed or edited. The GIS software manages a unique ID for each feature. The GIS software also populates values in the geometry columns for each feature, such as: *X, Y, length, perimeter,* and *area.* If the layer's projection is defined as having units in meters, the units in the table will also be meters. Users often add a column for other units needed for a report or table such as a column for acres, hectares or square kilometers.

The *shapefile* is a commonly exchanged GIS data format that has dominated the GIS scene.[12] They're quite awful in practice. The "database" portion of the shapefile is a constrained version of the archaic *dBASE* (DBF) database file format.

GRASS GIS uses a high-performance open source SQLite database by default, but it can also be connected to enterprise scale databases such as *PostgreSQL* with *PostGIS.* The *geopackage* is an open source GIS container that is also based on *SQLite.*[13] Geopackage is file or user based, while a PostgreSQL connection can be a multi-user enterprise solution.

More on these formats later in this chapter.

GIS Layers

For vector data, GIS layers are a collection of features that are grouped by a common theme. Raster data is also considered a GIS layer, which we'll get to shortly. An example of a GIS layer is a land cover layer. This is typically a polygon layer that defines the boundaries of land cover types (e.g., forest, water, developed). When we say theme, we mean a group of features that have commonality.

GIS layers are a collection of one to many features. Different geometry shapes are broken out into separate layers. For example, address locations could be stored in a parcel polygon, as a range on a line segment, or as a point. Each one has a different use case.

Some GIS software and data formats allow for mixed feature types within a layer and even if it is permitted it should be avoided because most GIS users are not expecting this behavior. If one finds themself in the peculiar situation of mixed geometry types, one should probably ask themselves why they are using mixed feature types. It might be an indicator that the features need to be separated into different layers that each have the same geometry type.

Another reason to use single feature types in a layer is that most software functions are built around specific operations for points, lines, or polygons; having them mixed in a file can confuse the process and intended output.

Following the prescribed geometric and thematic layer rules above makes working with GIS data logical and consistent for other users. As users create data, these thematic and geometric rules ensure that as GIS data is passed around, users are generally following the same conventions.

GIS Topic Categories

There usually isn't a problem when one GIS user is downloading and filing GIS layers into a catalog or organized filesystem. The moment there are multiple users accessing and creating GIS data, it becomes important to file the layers in a way that creates a library feel with browsable categories.

The *International Organization for Standards* (ISO) 19115 Topic Category list is

---

[12]https://wikipedia.org/wiki/Shapefile
[13]https://wikipedia.org/wiki/GeoPackage

a great way to group GIS layers.[14] We'll see how this is used in greater detail when we get into an enterprise GIS data library later. Users can always use these categories as a guideline and modify it to meet their own needs. It's a really great starting place, and actually end place, once you try to enhance or parse it out differently.

ISO 19115 GIS Topic Categories

Table 4.1: Categories and Descriptions

Code	Category Name	Description	Example Applications
001	farming	Rearing of animals and/or cultivation of plants	agriculture, crops, livestock
002	biota	Flora and/or fauna in natural environments	flora and fauna, ecology, wetlands, habitat
003	boundaries	Legal land descriptions	political and administrative boundaries
004	climatology-Meteorology-Atmosphere	Processes and phenomena of the atmosphere	processes and phenomena of the atmosphere
005	economy	Economic activities, conditions, and employment	business and economics
006	elevation	Height above or below the earth's surface	altitude, bathymetry, DEMs, slope, derived products
007	environment	Environmental resources, protection, and conservation	natural resources, pollution, impact assessment, monitoring, land analysis
008	geoscientific-Information	Information pertaining to the earth sciences	geology, minerals, earthquakes, landslides, volcanoes, soils, gravity, permafrost, hydrogeology, erosion
009	health	Health, health services, human ecology, and safety	disease, illness, factors affecting health, hygiene, substance abuse

---

[14]Source: `http://fgdc.gov/metadata/documents/MetadataQuickGuide.pdf`

Table 4.1 – continued from previous page

Code	Category Name	Description	Example Applications
010	imagery-BaseMaps-EarthCover	Base maps	land cover, topographic maps, imagery, annotations
011	intelligenceMilitary	Military bases, structures, activities	military bases, structures, activities
012	inlandWaters	Inland water features, drainage systems and characteristics	rivers, glaciers, lakes, water use plans, dams, currents, floods, water quality, hydrographic charts
013	location	Positional information and services	addresses, geodetic networks, control points, postal zones, place names
014	oceans	Features and characteristics of salt water bodies	tides, tidal waves, coastal information, reefs
015	planningCadastre	Information used for appropriate actions for future use of the land	land use maps, zoning maps, cadastral surveys, land ownership
016	society	Characteristics of society and culture	anthropology, archaeology, religion, demographics, crime and
017	structure	Man-made construction	architecture, buildings, museums, churches, factories, housing, monuments, shops, towers
018	transportation	Means and aids for conveying persons and/or goods	roads, airports, airstrips, shipping routes, tunnels, nautical charts, vehicle and vessel locations, aeronautical charts, railways, trails

Table 4.1 – continued from previous page

Code	Category Name	Description	Example Applications
019	utilities-Communication	Energy, water and waste systems, and communications infrastructure	hydroelectricity, geothermal, solar, and nuclear sources of energy, water purification and distribution, sewage collection and disposal, electrical and gas distribution, data communication, telecommunication, radio, communication networks

GIS Vector Overlays

Without sounding too fancy, there are geoprocessing tools, or functions, that can be applied to GIS layers. The features are, after all, just points, lines, and polygons. Not only can we do things like buffer or offset, from these features, but we can also overlay two layers and perform analysis. There are different functions that can combine or exclude GIS vector features when they are compared. These geoprocessing tools are the basic GIS functions users need, especially the combination of functions.

The clip function can be thought of as a cookie cutter. The underlying 'dough' is the GIS layer we want to clip a smaller piece from. This smaller piece is usually a project area or boundary. A clip function can be used to create a land cover summary for a parcel. The parcel boundary can be used to clip data from the land cover layer. When this output layer is summarized, a table can be created that shows a summary of the land cover types in the boundary. If we have good data topology, it will add up exactly to the area of the project parcel boundary.

Along with clip, buffer and erase are commonly used geoprocessing tools. In an example using a solar field, let's assume there is a constraint that wetlands and areas within 50 feet of a wetland cannot be developed or impacted. First, we buffer the property 50 feet and then clip out wetlands from the wetland layer. We buffer the property to clip and consider any off-site wetlands that might buffer into the property. Then we buffer the clipped wetland layer 50 feet. This buffer includes the wetland plus the buffer, so we can use this buffer file to erase areas from our starting project boundary. The output layer has polygon areas that are not in, or within, 50 feet of a wetland on the parcel of interest.

A common overlay for GIS layers is the union. There are variations of this, such as intersect and difference, but the union is the general one used in wide application. Building on our solar example above, we have an output of polygons that are 50 feet away from wetlands. Now we want to identify agriculture fields or grassland areas. The output layer containing polygons outside wetland buffer areas can be used to clip the land cover layer. The resulting layer can be queried for land cover areas that meet the criteria for grassland or agriculture. In practice, unions are often performed on layers and then users can query the combined attributes in the output GIS layer. Extremely useful and a cornerstone of GIS analysis.

4.7 Raster Data

Raster data is another common type of GIS data that users encounter. A picture is a type of raster data that we all are familiar with, it has colorized pixels and a resolution (pixel per some unit of length measure). Another type of raster data users commonly encounter is a *Digital Elevation Model* (DEM). These store elevation values in the pixel instead of a color, although the elevations can be colorized. Vector data is discrete in that it is a specific location, line, or area. One of the main benefits of raster data is that it models continuous values over an entire area (without gaps or holes), and because of this feature, raster data models can solve some interesting GIS problems.

Raster data has a gridded data structure, established by rows and columns. The other defining characteristic is that raster data has a cell size, and therefore it has a resolution. When working with raster data, the cell size is perhaps the biggest consideration. Make the cell size too big and details like drainage ditches may be missed. Make the cell size too small and it may be too large to process. A common trade off with raster data is the cell size and the extent, together they form the number of cells. When memory limits are exceeded, the cell size needs increased, or the extent needs decreased. If the cell size needs to be maintained, the data will need to be processed in segments and patched together to get around resource limitations.

Topography, or surface elevation, is an excellent dataset to illustrate the difference between raster and vector data. One common way to represent topography is with topographic contour lines. These are *isolines*, lines with the same elevation value. Contour lines are discrete, in that they only provide elevation values where the lines are; no elevation values are known between the lines because there is no data. A raster DEM on the other hand, has continuous elevation values for the entire area. Every pixel in the raster has an elevation value. Each can be useful to illustrate elevations for different scenarios, but for modeling, a continuous surface is always preferred. Modeling hydrologic features or line-of-site visibility always requires raster data for the analysis.

Another common type of raster data that GIS users encounter is remotely sensed satellite data. There are many different satellites and sensors that capture data from space. One well-known land cover layer for the United States is the *National Land Cover Database* (NLCD). This GIS layer classifies land cover types from satellite data into meaningful values such as developed areas, wetlands and forests. The data is also available for different years, which is useful for historical analysis. Oftentimes, this raster data is converted to polygon data so that regular vector overlays can be used on the data such as a clip and acre summary.

More About Elevation Models (DSM, DTM, CHM)

While there are different variations, almost all DEMs distributed are bare earth DEMs. Bare earth DEMs represent the elevation of the earth's surface, without objects on it. Less common are *Digital Terrain Models* (DTMs), which usually have integrated features such as streams or roads. Not that these features are not already in a DEM, but a DTM usually provides other "help" to the DEM by way of explicitly defining these features.

Another type of elevation model is the *Digital Surface Models* (DSM). These elevation models have ground elevations as well as features above the ground such as: houses, electric transmission wires and trees. DSMs are very useful for line-of-sight modeling, because they consider features such as buildings, which might obstruct the observer's viewshed. A *Canopy Height Model* (CHM) is another example of an elevation model that is used to model the tops of tree canopy; a rather useful model for forestry.

Null Values

The raster data structure has a special value called the *null value*. Sometimes we see this value in a vector attribute table, but the null values have no impact when the vector layers are used in geoprocessing. With raster data, if null values are overlaid, the null cells will not be in the output. This may seem odd, but null means that the cell has an unknown value. There is no data for the cell so operations can't be done to it other than converting the null value to a number.

Raster maps are often combined mathematically, and when we look at it this way, it makes more sense that null values are not put in a raster output. Any number that does math with a null value outputs a null value. In a raster data structure, every cell must exist, but may not have a value. Null values are not rendered when a raster layer is drawn. If math needs to be done with the raster layer while considering null areas, the null values can be converted to zero or another number that will allow the operation to consider those areas.

Raster Editing

Another difference between raster and vector data is how they are edited. Raster data is usually created from something, a sensor, a model or analysis. It's also a useful way to illustrate data along a continuous color ramp. When it comes to editing raster maps, it is usually done by creating the shapes as vector data, and then converting the vector features into raster data. Editing point, line, and polygon shapes and then converting them, is much easier than trying to edit something like a photograph with square cells and values.

An example of a raster edit would be adding a weir to a DEM. A weir is a wall-like feature that holds water back. When a weir is added to a stream, the water can flood beyond the stream banks until it goes over the weir at a given elevation. GIS can model this action by creating a weir feature in the DEM. One might think a line would suffice for the weir feature, but it's best to add some width (polygon) to be sure and capture diagonal cells that can be tricky. After the weir is created, convert it to a raster feature. Then in the DEM, calculate the area of the weir to be the desired water elevation.

Remember to be mindful of resolution when working with rasters. If you digitize a one foot wide weir but are using 20 foot cells, guess what size weir you end up with? The same consideration should be applied to stream carving and other data revisions that work with raster data.

Raster Calculator

One of the most common tools, in any GIS software, is the concept of a *raster calculator*. Raster layers have a projection and a cell size. Overlapping raster maps can be treated like a math expression. These expressions have an algebra-like syntax. Sometimes the operations are referred to as *map algebra*. It's pretty cool once you get the hang of it.

An example would be converting a DEM from meters to feet. A simple conversion would be: `DEM_in_feet=DEM_in_meters * 3.28084`. This expression multiplies each cell in the raster `DEM_in_meters` by `3.28084` and creates an output raster with the new values, `DEM_in_feet`. Raster calculators can use conditional statements that can slice and dice raster layers in many interesting and useful ways.

Site Suitability Using Raster Overlays

There is one type of raster analysis in GIS that is very powerful, known as site suitability. Site suitability analysis uses the continuous nature of raster layers to assign suitability scores to the study area. It considers many GIS layers and outputs a raster layer that ranks locations based on how close they match the desired needs. One cool raster function calculates distances from features across the entire raster extent. As cells get further away from the source features, the distance is calculated and stored in each cell.

For example, suppose you want to build a solar site within 2,000 feet of a high voltage transmission line and have a preference to be as close to the transmission line as possible. After running the distance function on the power lines, the output raster cells will contain the distances to the nearest transmission line. This layer can be combined with other suitability layers, such as distance from residential areas, creating an overall suitability ranking that considers both variables. The raster cells that are closest to transmission lines but furthest from high density residential will have the highest score as they are more suitable.

The same analysis isn't possible with vector data. Is it clear why? Vector data is very useful for mapping constraints and eliminating them from a project area; more of a binary approach. So in the case of wetlands, water bodies, and streams, let's assume they cannot be developed or impacted within 30 feet. We can use our familiar buffer function to establish these undisturbed areas, and then the erase function can remove these areas from the project study area. Note the difference between removing features and ranking them for suitability.

Vector or Raster?

How do we know when to use vector or raster data? It may not be clear at first, but with experience comes the knowledge of which data model to use. Vector data models discrete features while raster data models continuous phenomena. Most mapping situations require the use of vector data, while modeling and analysis often utilize raster data. There are exceptions to this, of course. In general, if solving the problem cannot be done with points, lines, or polygons, then a raster-based approach is probably the solution.

4.8   GIS File Formats

Shapefile

The most common data format on the Internet is the *shapefile*, created by that dominant commercial GIS company.[15] Shapefiles are not an open source format, but over time the details have been released in a white paper that has allowed developers to read and write the format in other software packages. This file format was created in the early 1990s and the fact that we're still using them 30 years later is a bit alarming. Shapefiles only have single precision to store X,Y coordinates and the "database" has severe limitations, like 13 character field names. The latter has proven to be very challenging with regards to working with new GIS data, the often truncated fields make it very difficult to know what the attributes are after a conversion.

The name is misleading because it is not one file but many files that work together in a GIS. There are three essential shapefile extensions: .shp, .dbf and .shx. The .shp file holds the geometry, the .dbf file holds the database information and the .shx file is an index file linking the two. Over the years, an absurd amount of helper, or sidecar, files have been added such as: .prj, .sbn, .sbx, .fbm, .fbx, .ain, .aih, .ixs, .mxs, .atx, .xml, .cpg, .qix.

Two of the familiar file extensions are .prj and .xml. The .prj file holds projection information and the .xml file holds metadata about the GIS shapefile. One shapefile today usually consists of about seven files.

Shapefiles are unfortunately common but they need to be discontinued. From a user perspective, it's much better to have GIS layers in one file or database. Having a directory full of shapfiles, with 7 files per GIS layer, can get quite messy. Not to mention the technology limitaions found in shapfiles. There's a great webpage dedicated to this idea that "Shapefiles must die!"[16]

Geodatabase

In the commercial GIS world, the most commonly used GIS file format is the *geodatabase*. APIs have been developed so users can read data from the Geodatabase, but initially they were closed and only ArcMap could read and write to them. Imagine the frustration as an open source GIS user, trying to use GIS data that was in a locked format.

---

[15]https://en.wikipedia.org/wiki/Shapefile
[16]http://switchfromshapefile.org/

After ARC/INFO coverages died, there was a Microsoft Access geodatabase called the *personal geodatabase*. This bad idea of stuffing large amounts of GIS data into an Access database didn't last long. It was a brief time (a few years?) with the personal geodatabase and its whopping 2GB file limitation and then the *file-based geodatabase* came next and is commonly used today. The *enterprise geodatabase* embeds the structure of a file based geodatabase into an RDBMS. This extends the file geodatabase further by using RDBMS technology.

Geopackage

Geopackage is an open GIS file format that is based on a SQLite database container.[17] This file format is used very frequently with open source GIS desktop software. More US government agencies are also providing public data in this format. A replacement for the shapefile was long overdue and this single file, open database is the perfect evolutionary step.[18] There really is a desperate need to replace shapefiles with a modern, efficient and open GIS data format.

PostgreSQL with PostGIS Spatial Extender

The geopackage satisfies the single user, but if an enterprise solution is needed, look no further than the open source combination of PostgreSQL and Post-GIS. PostGIS is a powerful spatial database extender that can be used with the RDBMS PostgreSQL. PostGIS provides spatial queries against a database without the need for a graphical component. Both QGIS and ArcGIS can read data from a PostGIS database, which will be explored in a later chapter through a GIS data library. It's important to know that an RDBMS is a network service, not a file. This means the data is accessed by a specific IP and port number and is not bound by office locations; the data can be shared across the enterprise.

Raster Data

Raster data is a little less exciting and there is one file format that is widely used: the *GeoTiff*. The GeoTiff is a spatially enabled *TIFF* image and can have a projection and coordinate system assigned to it. GeoTiffs have the same extension as a TIFF image, .tif. This versatile image format can store integer and decimal numbers and offers some compression to minimize size. Other

---

[17]https://en.wikipedia.org/wiki/GeoPackage

[18]Geopackage website: http://www.geopackage.org/

common formats are the *Erdas Imagine* format, .img and the *HDF* format that some satellite data is stored in. GRASS GIS uses its own raster format, which QGIS can read and display.

A Few Notes on GIS Data

Open source software is based on the idea that information should be open and transparent. It follows that GIS data should be in a format that is open and accessible, so that any GIS software can read the data. GIS data is essential to scientific research and studies that are occurring around the globe. Science also follows the belief that data should be open and able to be reviewed by peers. Since GIS is science data, the formats that store this data should be accessible and in an open format. At a minimum, reading the data is an absolute.

To be clear to new users, users generally don't get to pick a GIS file format; they use the format that has the best features in the software they're using. Using commercial GIS, the geodatabase is the only format that supports topology. In GRASS GIS, the internal vector format can be used or it can be connected to a PostGIS spatial database. In QGIS, the geopackage is the best choice for a file-based GIS data container. The point is, an ArcMap user wouldn't use a geopackage. They could, but a QGIS user can't use a geodatabase in the same way a user would utilize a geopackage.

4.9   Cartography: The Art and Science of Making Maps

Map Making

The last part of GIS 101 in university deals with maps or specifically *cartography*. Some users love it, some users hate it and others fall somewhere in between. Cartography is an essential part of GIS 101 because it shows or illustrates the data to end users. More recently, GIS maps have evolved to a digital format of PDFs or web formats delivered to user desktops or mobile devices. Either way, maps have a few basic requirements.

Cartography is the art and science of making maps. It's artsy in the colors and symbols and it's sciency in the projections and geometry. It is an old discipline that goes back 20,000 years. Cartography can be a role performed by itself, where the user is not creating or analyzing data, just making maps from existing data. Cartography involves a lot of symbols, colors and annotations. Users who have a knack for color and aesthetics really excel at this part of GIS.

All Maps Lie

"Why do these innocent maps lie?", one may ask. It's in their nature. How dare you! In all fairness, taking something round in 3D space and presenting it as a flat 2D object is not possible without consequences. A way to visualize this issue is to picture an orange cut in half with just the outer skin left intact. Now imagine trying to show the shape of that half-orange on a flat surface. The flattening of that round object into two-dimensional space forces the orange skin to become distorted or ripped.

There are many different examples of global map projections discussed online.[19] Something that may not be initially obvious is that the smaller the area being mapped from that orange peel, the less distortion there is in flattening it. This is a great illustration why small project areas don't have to deal with distortion issues as much as larger extents do. Regardless, projections and coordinate systems should always be at the forefront of any GIS operator's mind.

Projections take a curved object, the Earth's surface, and use mathematical equations to display the objects on a map. When projecting GIS data, something has to be distorted. GIS projections exist to minimize impacts to shapes using different projection techniques and types: conic tangent, conic secant, cylindrical planar, and polar. They each touch the earth's surface as a plane using different math equations. This is where the controversial statement "all maps lie" comes from, the fact that all maps have to impact the data in some way; something always get stretched, squashed, or rotated.

Data Deception

Projections usually aren't used to intentionally deceive map users, but symbology can be used to display a map in a favorable, or unfavorable, way. For example, when wetlands are assessed for a project, they are given a score by professional ecologists. The score ranks the wetland from 0 to 1 based on a bunch of quality and functional characteristics. How should the score be displayed qualitatively? Where is the break for "good" and "bad" wetlands?

Given two opposing groups, a developer and a nature conservancy organization, there are different interests in how to display the wetland quality data. With hundreds of wetlands to display, the developer may categorize the wetlands as: *0-0.8 Low Quality, 0.81-0.91 Medium Quality*, and *0.92 - 1 High Quality*. Categorizing the scores in this way would identify wetlands lower than a 0.81 as low quality, but is it correct to do? Maybe the nature conservancy group

---

[19]https://loc8.cc/osg/projections

makes their own map from the data, flips the high and low values, and then sends the map to media outlets. Who's right?

The data is the data and unfortunately some of it is open to interpretation when it comes to symbolization. When GIS data is submitted to a public agency for a permit or review, it becomes public data. GIS data doesn't have to be falsified through editing the data, it can be done deceptively through symbolization and other map display techniques. Overemphasizing one area can be done with shading or simply by minimizing where attention is drawn, showing what the cartographer wants to focus on.

These components of cartography are very important. Even aside from how data is manipulated or displayed, cartography requires a good eye for art and how data should be presented. Color ramps, symbol sizes, and colors—there are so many ways to be creative with GIS maps that don't involve working with or building the data. Most cartography has migrated to online maps, but the same concepts still apply. From experience, don't be surprised if you do an awesome analysis and the only comment is to change purple to lavender.

Cartographic Elements

A map needs to have several components on it to make it a useful map. These items are: the map of features or base layers, scale bar, north arrow, map title, and legend. The scale bar tells the user how many units on the map equal real units on the ground. The north arrow lets readers know the orientation of the map. The title tells the reader the purpose of the map in a few short lines. The legend tells the reader what the colors and symbols on the map represent. In recent years, the "Legend" label has been dropped on maps. Other useful items to place subtly on the page: the user name, filename, file path, and date.

The Angry Color Red

It's more than likely users creating maps will encounter that person yelling, "Don't use red! Red is an angry color!". Not really sure where this came from or who's teaching this, but it needs to go. In theory, red may represent or be associated with anger, but in reality, red is one of the best contrasting colors for maps with an aerial background; although yellow works good too. People will insist that red is an angry color and not to use it, especially when trying to appeal to regulatory agencies who are making decisions based on maps. For what it's worth, in over 25 years, can't recall anyone getting mad over red on a map.

4.10   The Wrap-Up

Gaining an academic background in GIS was extremely valuable. It set the foundation for GIS concepts and applications, as well as providing a solid knowledge base about raster and vector data. Points, lines, and polygons in GIS are pretty straightforward to understand and work with, while projections and topology can be a little more abstract. It comes with experience, so just keep plugging away and soon all the pieces will come together.

Some users will work with GIS through buttons and have a simple understanding about layers and how to display them on a map. Other users will expand their knowledge and skills to mix raster and vector data layers into complex models and analysis. The great part about GIS is that you can use it as your primary area of expertise, or use it to help you with your primary area of expertise. The level of involvement is as deep as you want to go.

While in university, it was difficult to picture what the end result would be. There was already an interest in hydrology and hydrogeology and several courses had been taken. Now that GIS had been encountered too, there was great excitement in the prospect of using GIS for hydrologic analysis. We would pursue this and many other GIS adventures on this journey. Buckle up. Actually, it's a GIS journey, so it probably won't get all that wild, but hopefully it will get very interesting.

That's it. GIS 101 from Seibel University; consider yourself an official graduate.

5. Graduation to Entry-level Work

5.1 Still No OSS

No open source yet? How can this be? Unfortunately, open source software has not crossed paths with this journey yet. We got a little sample of open source software in chapter two, but nothing really tasty. We're almost there, just a little more. Unfortunately, or fortunately as seen later, the seas of this journey have to get rough before they get better.

A congratulations is in order for making it this far. The last chapter was a necessity and it really had to be done. What GIS is and how the data works are the building blocks to bigger and better things. Eventually, we'll get to the place where we are predicting stream locations, modeling stream banks, modeling drainage networks and computing lake volumes. These applications are not run-of-the mill outputs from canned buttons; they require understanding about what's going on under the hood.

5.2 Suggestion on How to Start

Okay, so you took some classes or read a book and now wonder where to begin. After "What is GIS?", comes "How does it work?", then "Where do I start?". The answer to this question depends on the answer to the question: What do you like to do? The point of answering the question with a question, is to refocus the context around the individual. We don't just "use GIS software"; we have to apply it to something.

The widespread use and various applications of GIS create an environment with a wide array of characters playing different roles in it. We have engineers, artists, nerds, data dudes and dudettes, scientists, hardcore users and informal users, all mixed up a huge community. We've got it all. As described in previous sections, there are many different roles to fill in the field of GIS.

The key is to pick something that matches what you enjoy doing. Do you like cartography, analysis, data collection, data creation, science-based data exploration, marketing, systems, project management or GIS management? Just pick one and go for it, you can always change it up later. This particular journey had a foundation of technical knowledge in computers and an Environmental Studies education. These two are a great combination; what does your combined education, experience and interest yield?

5.3   Entry-level Work

Ah yes, the good old days of entry level work. In the GIS field, starting out can be quite discouraging because most of a project involves creating data or collecting it. Don't be surprised if you click your way to one million mouse clicks. Data creation in GIS can be very tedious, tiresome and even damaging to one's body due to the repetitive nature of the mouse clicks and keyboarding. Mass digitizing is often delegated to entry level staff because it is relatively simple and is usually a bulk of the work needing done. In many businesses, there is a lower cost associated with having an entry level worker do the basic tasks.

The entry level position for GIS is the technician. Fortunately, this journey had a short stay at the technician level, so it won't be too boring. A GIS technician's primary tasks are working with data and making maps. The biggest effort and amount of time in GIS projects is creating data. One estimate is that data creation is 80% of a project cost, which seems about right based on experience. This can involve digitizing a lot of data; so much data that your eyes see land use and topo lines when they are closed. It can get that bad.

In addition to digitizing a seemingly infinite number of geometry shapes, there can also be a high volume of database entry. The attributes for features in GIS need to be populated with values and some datasets can easily be in the thousands of features. For example, someone working in a property appraiser's office may spend the majority of their time changing the database attributes. This situation would probably have more database attribute changes than parcel boundary geometry changes. When land is sold, the boundary - or polygon geometry - stays the same but the parcel owner's database information changes. GIS technician work in this position might mean a lot of text updates to the database.

5.4   The Work-Experience Paradox

The best way to get any job - aside from knowing the right people - is to have experience. The entry level paradox: How can the requirement for experience be met if the person has never done the work before? Experience is usually the primary prerequisite, independent of one's educational focus or non-related experience.

Many people encountered GIS and gained their experience at college or university. GIS is usually encountered as a support software to a primary subject such as environmental or geography. Learning GIS in an educational setting provides for a solid background in theory and a bonus benefit to interact with the professor and ask questions. With the explosion of GIS at the turn of the century, GIS in the educational systems expanded to include online courses and GIS certificates.

There is another way to gain experience and that is to learn it on your own. Perhaps the first thought is that the software is expensive and it would be too daunting of a task. Free and open source software has a distinct advantage in this area, being that the software can be downloaded for free. The good news is that QGIS is an amazing desktop GIS and the community surrounding the software is very supportive and welcoming. There have been many instructional videos and manuals created by the community for new or advanced users. The concepts in GIS dont change with software, so learning one software will translate to learning 80% of any GIS software. Users can collect GPS data with their phones and do a simple data import to QGIS. This is a great way to gain exposure to GIS software without any expenses.

Another idea is to use free trials. To gain experience with commercial GIS software, companies will allow limited time-trials with their software. This applies to the core software and extensions. They also offer some free technical courses and certifications to get started. This can help new users get acquainted with the software without spending any money.

Internships are also a good way to get involved with GIS and add some validity to the experience. Unfortunately, creating GIS vector data often requires a lot of manual work. Don't hesitate to offer some free help to a company to get experience using GIS software. It doesn't take long to get familiar with all the basic tasks and functions that are required for most GIS requests. A summer internship can be really beneficial for software experience and using it in a professional capacity.

This journey goes for 27 years and has its start in university, a hop to summer and short jobs, then 25 years in the environmental consulting industry. It was in the environmental consulting industry where GIS would be applied to many real world applications. Like all good things and times coming to an end, the days in university were over and it was time to get to work using GIS. Not without one last fun summer at the beach.

5.5   Wish All Jobs Could Be Like This One

Life at the Coast

The catapult into GIS came during the last semester at university. Oddly, the cartography course using ArcView 3.x was taken after the Intro to GIS and Advanced GIS course. Perhaps a signal for the unorthodox approaches in the upcoming journey. During the cartography course, the professor asked about working with him for the summer. This would be another great influence in the journey, the first of a few supervisors to shape the path. A room was rented on the waterfront and the last, maybe best, summer was about to unfold.

In addition to the great job, the house on the bay wasn't awesome but its location was. It was a few minutes from the beach for surfing, the icing on the cake. The landlady was cool and let us user her small boat for fishing and and sight-seeing. Waking up to the sounds of the shore and seagulls, and the smell of the beach was really hard to beat. Other times we would drive the boat up to the restaurant and have some food and drinks with the professor who undoubtedly made his best hire from university. Sorry to diverge, but sometimes it's just great to reflect back on days that made you really smile. Enjoy these early days before things get too seriou and don't forget to have some fun along the way.

The First Job

Ah, the job. The job involved taking water samples in the ocean and doing some GIS work. Who could forget the summer days at the health department taking water quality samples? We would be out on the beach in the Jeep by 6:00 am and catch the sunrise over the ocean. It was really tough work at times. Sometimes we would swim out into the waves, collect our sample, then body surf the waves back into the shore. Of course bodysurfing was just an extra high-level skill we applied to the job to maximize efficiency in sample collection. There was also the county boat that was used to collect samples in the bay areas and an offshore ocean boat trip to check the outfall pipe.

What does this have to do with GIS? Nothing. It was just a really fun component of the job. We did do some neat GIS work though. After our morning data collection at the southern New Jersey beaches, we would work on writing an AML (Arc Macro Language) script. The AML would take a project area and analyze it for septic tank suitability, specifically the drainfield. Turns out the high school programming courses were going to be quite beneficial going forward, writing scripts to automate GIS tasks.

That was it. The first summer out of university had come and gone. Some GIS was worked on while having some final fun days at the beach. The septic suitability program we made even made it to a presentation for the county. Not a bad start and it was all owed to my first guide in the GIS world. It was here it was learned to have fun at what you are doing and also get the business part done. He also taught about the value in side-hustle with GIS work; good information to know. This was the first of two professors encountered on this journey and would become a friend. The first shaped my career and the latter shaped my life. Be thankful for the people you meet along your journey; each has a unique contribution to the curves and straight-aways in the trip.

Leaderless Journey

While there may be references to a mentor or two, to be clear, it wasn't like there was a coach who could lead and teach about all the special things GIS can do. Or even have someone around to answer simple to complex questions. Everything, and indeed everything, was driven from within; a raging curiosity to explore and try new applications of GIS technology. All these things had to be figured out without a mentor, and that was just fine. Yes, there were people early in the journey who provided some basic knowledge, but it really wasn't anything that hadn't been learned in university courses anyway.

Open source software provides a near infinite source of curious things to explore and a very helpful community to interact with. Open source software is not only free in cost, but provides freedom in choice and exploration. Open source involvement and in depth knowledge is something the individual has to pursue and have a passion for and there is no mentor required.

5.6   Getting Experience Any Way Possible

The summer had come to an end and so did the summer job. There were no full-time positions open at the health department, so it was time to move on to the next job. Many jobs are obtained by knowing the right person. While this next step was no exception, it was the only instance. The professor knew someone who had an opening at the New Jersey Department of Environmental Protection (NJDEP) for a GPS Technician. After waving goodbye to the beach and the professor, the next step was entry level GPS work. No open source encountered yet, bummer.

The next supervisor at the NJDEP was really cool; 2/2 so far! The GPS job had a fun component too, although nothing could match the bar set by the county job, or ever would. This new job entailed driving around south Jersey in a 1989 VW Golf, logging facility locations that had chemical contents of some kind. The information collected could be used for fire response and other organizations that need to know the potential dangers during response situations. The GPS work was definitely going to be a temporary thing while seeking a job that was focused in GIS.

There were a lot of beautiful places visited along the way while logging facility locations. The process was to drive to the address, verify the building and log a GPS point. Aside from the great south Jersey views, it was as boring as it sounds. Prior to logging the GPS point, a large magnetic antenna was mounted to the top of the car to increase accuracy. Sometimes the antenna was left on top of the car, unintentionally, between facility visits. Shout out to the 1990s Trimble antennas. These antennas turned out to be quite durable, as they sometimes would fall off the roof and bounce on the road, still cabled to the GPS unit. After collecting the data, it would be post-processed with ground towers and then the data was periodically driven to the headquarters in Trenton, NJ. The points were then put into a GIS layer as part of NJDEP facility inventory.

Remember the tip from a friend in college about GIS becoming a hot field back in college? Well, it turned out to be a good prediction afterall. After looking for an entry level position using GIS, two job offers were proposed: one from a private environmental consulting company and the other from a quasi-public utility company. The "quasi" part was never fully explained. This was during the era of sending cover letters and resumes out via fax machines. Ah, the good old days, eh?

5.7   Public Works: Not for Everyone

The public work avenue was chosen and the hiring manager at the environmental consulting company was really disappointed. It's okay, we'd be back. He had recently learned about GIS and was excited to have someone with this background in his permitting department because GIS skills were not all that common at the time. After starting the quasi-government job, it wasn't long before realizing this wasn't going to work out, at all. There was a burning desire to do something more.

The job was mostly about field data collection. Different water network features were to be mapped with GPS: stormwater inlets, manhole covers, valves and related infrastructure. After field data collection, the data was brought back to the office for differential correction using base stations. After post-processing, the GPS data was imported into GIS and appended to existing GIS inventory layers. This was the first time, and last time, PC ARC/INFO on MS-DOS would be encountered. PC ARC/INFO had limited functionality compared to UNIX ARC/INFO, not even sure about the main differences, but it did have single precision.

My new supervisor was another cool character - 3/3, nice! He was probably the first computer nerd met along the way, of which there was a bonding and enjoyment for someone of this similar technical nature. Nerd is, of course, used in the most caring and self-identifying way. He was a SCUBA instructor and we got trained and certified by him in a rock quarry in Pennsylvania.

It was only 1997 and he was a constant propagandist calling for how GIS was going to be everywhere and everything. He insisted that we were at the ground level of a technology that was going to explode in the future. It sure did and in the US it got ensnared in commercial GIS software and their practices.

This supervisor was great. He found a second job for a graveyard shift and asked if there was any interest to join him. An interesting side-hustle again, although the work was incredibly boring. It was in this highly secured work environment and involved data entry for addresses in a 911 database. Ah, the joys of data entry. This one involved working through the night shift on a Friday night into early Saturday morning. The drive home was tiresome and dangerous, nearly falling asleep at the wheel. It was a good thing that we were let go after only the first shift because the driving was unsafe and the work was unbelievably boring - gobs of data entry.

Back at the GPS data collection for the water network, we chugged on, day after day, doing the same thing. Collect field data, process, append. The people were fun, but it was so mundane, more than could be tolerated. After three short months, it was realized that this type of involvement with GIS would not suffice. Challenges were needed. Conundrums needed to be solved. Applications of GIS that required thinking and problem solving were calling and it was clear that a new home must be found. The thought occurred, what would the environmental job have been like?

5.8   An Emerging Field and Envisioning More

The more this was pondered, the more thoughts went back to the other job offer. What kind of work might be performed at the environmental consulting firm? It seemed likely that a B.S. in Environmental Studies, accompanied with core focus in hydrology and GIS and might be a better fit at an environmental consulting firm. Plus, it was hard to forget how excited the other hiring manager was to introduce GIS to the company he worked for. He tried to learn and use GIS himself, but found that there was more than meets the eye. He wanted someone with expertise who could hit the ground running.

The environmental firm had a lot of appealing aspects to it, with regard to the professional application of GIS. Pride was swallowed and a call was made to the previous hiring manager. He was thrilled at this decision and a job offer was made. After joining this environmental company, the journey really accelerates and aligns exactly with a GIS Professional job. This is where the power of GIS is unleashed for the first time with great success. The decision would turn out to be the right one, and should be an example to others to pick what you think interests you. Don't be afraid to change your mind and explore options while you can; this advice doesn't seem as necessary to the current day job-hopping phonmenea, but is still good advice. There are all kinds of job types and the longer one stays at a place, the harder it is to leave. Like that state hotel song.

We would eventually leave the water authority and go to the environmental consulting firm. In this next stop, we're moving to a new network environment that uses Windows NT coupled with that commercial GIS software. In this setup, Windows NT would be the data server and the client workstation would be running ARC/INFO. Goodbye UNIX, at least for a while, and hello Windows.

This commercial road was going to be bumpy, with a typical Wily E. Coyote road disappearance and a major crash and burn at the end. Fortunately, we're more like the Roadrunner and escape to the next destination unharmed. The Wily E. Coyote reference is a harbinger for the woes and suffering with commercial GIS software to come. In fact, this part of the journey was so exasperating, it escalated to the point of having to make a crucial decision whether or not to eject from this journey altogether. Working at the environmental consulting firm is where things started to fall apart. The worst of it peaking a few years later in 2002.

5.9　Historical Perspectives on GIS

UNIX to Windows - Scoop Up the Cash

Pausing between the public works and private consulting firm, now's a good time to some historical background on commercial GIS and its first transition. The commercial GIS software ARC/INFO on UNIX was very reliable and stable. Whether it's worth the cost or not, you get to be the judge of that when this is all over. ARC/INFO Workstation was the analytical commercial GIS software that utilized coverages and topology for vector data. There were no crashes experienced with the software while using it in university on UNIX. The same goes for the GIS work on UNIX at the health department; solid and reliable performance. If something was amiss at the end of an operation or analysis, it was without a doubt the user to blame. In a few short years, this would completely reverse to, right or wrong, immediately blame the software.

At this point in the journey, it was 1997 and that commercial software company ported ARC/INFO Workstation on UNIX to the Windows NT operating system. There was a PC ARC/INFO, but it had limited functionality and wasnt that prevalent. Much like on UNIX, ARC/INFO Workstation on Windows NT was not considered user friendly with its command line interface and lack of GUI. Okay, yes, there was some archaic GUI that could be launched in Arc Workstation sessions, but it was awful and nobody used it. The command line was much faster anyhow.

Two Products, Two Languages

There was also this commercial desktop software called ArcView by the same company. Why do we care about deprecated ArcView 3.x? We don't, other than some interesting situations in the history of this GIS evolution. ArcView 3.x was written in a proprietary language called AVENUE. ARC/INFO, as best confirmed, was written in C+ and FORTRAN. The two products could not be integrated since they did not share the same programming language.

ArcView 3.x was primarily used for cartography because making maps with ARC/INFO ArcPlot required programming in AML (Arc Macro Language) - a scripting language for ARC/INFO Workstation. For large jobs requiring dynamic labeling, ArcPlot was awesome; but for a simple map, ArcView was highly preferred with its GUI driven cartography. ArcView could view coverages, edit shapefiles and do some analysis. Although extremely popular, ArcView 3.x was always a second-rate GIS with its shapefile GIS layers. The shapefiles consisted of 3-8+ clumsy files and a non-topological geometry data format that always reduced ArcView to a poor GIS.

Users Flock to ArcView 3.X

Why even mention this ancient history? ArcView 3.x is the crucial pivot where this commercial software giant gained mainstream traction. This shift from UNIX to the Windows operating systems expanded the commercial software giant's reach. Windows was the dominant desktop software and cheaper on all accounts: hardware, software and administrators. This move to a more mainstream operating system began the tidal wave of GIS exposure and growth; eventually drowning UNIX altogether.

UNIX and Windows are very different operating systems in terms of design. Without starting a heated debate or argument, UNIX is more reliable and has had a much longer time fulfilling networking needs of large organizations. We're accepting this simple fact and moving on. The high computing power of UNIX machines coupled with reliable networking, made UNIX an ideal operating system for ARC/INFO. These GIS files had many files that made up a GIS layer and accessing them wasn't as straightforward as a single document. Common GIS network configurations were a data server and workstations; this allows for best practices of backups and redundancy for centralized data.

By the late 1990s, Windows (95 or variation of) was on every desktop so abandoning the original user base of UNIX users was no problem financially. The Windows platform offered a lot of sales opportunities. UNIX was cost prohibitive but every company or organization was running Windows servers and clients. Windows NT was the first Windows operating system that could run ARC/INFO, but porting ARC/INFO UNIX to Windows NT came with major issues.

5.10 Environmental Consulting: Finally a Good Match

Back to the journey, it was time to leave the water utility work and head to the next stop, an environmental consulting company in North Jersey. Cool boss #4. He was the supervisor of a permitting department and wanted to use GIS in his department.

GIS was relatively new to most smaller companies so there weren't titles for GIS employees. In this case, the formal title was a hydrogeologist, but the work was solely using GIS software. This shows how new GIS was to industry in the late 1990s. The first assignment at this new place of employment was to develop a Risk Based Corrective Action (RBCA) analysis in GIS for a contaminated site. There was a formal process for RBCA, but the current task was to use GIS to display data and create an interactive GIS application about the site. The project was a really good application of GIS, as it is today with Phase I and Phase II assessments.

GIS Site Assessment Work

The work involved creating and using data for a visual assessment of a contaminated property. There was a public complaint about the contamination from an adjacent or nearby salon. Parcel data was georeferenced and added to the project to show property owners near the site. We mapped the parcel in great detail with an aerial photograph and topographic contours to illustrate groundwater flow direction. At the time, there was some GIS data available, but public GIS data was not as abundant as it is today. Most of the project data had to be digitized and created. The groundwater modeler provided a modeled contaminant plume and it was georeferenced in GIS and overlaid with the site. The map and information came alive once they were all together in a GIS. It was an excellent way to visualize all the relevant layers and conditions of the site.

There was a lot of data collection and georeferencing into GIS, but interestingly there wasn't really any analysis. This experience would highlight the power of visual illustrations and presentations using GIS software. GIS software can be a great presentation tool. It does an excellent job at clearly presenting many interesting data layers for a project. Layers can be symbolized and turned on and off in such a way to show their combined presence on the site. The map can be zoomed in and out while data can be clicked on to get information from the database. It's a great way to keep an audience engaged and handle questions on the fly.

The data building part was going just fine, but the boss wanted a stripped down ArcView 3.x that only had the buttons we needed. Can't really recall why, but perhaps it was to add a look of customization and simplification. Customizing ArcView required learning the Avenue programming language. All the buttons were removed except a dozen for fixed zooms and queries. AVENUE was a new language to use, but the programming experience in the past made for a relatively easy adaptation to the code.

This is a good spot to suggest another GIS tip going through a GIS journey. If you have any aptitude towards programming, take some courses or read a book. It can be extremely useful to know how to program when working with GIS software. It's not to say the person needs to be a master programmer, but just knowing some scripting with the basics of variables and loops. If nothing else, learn some Python. It's a very popular programming language both inside GIS software and out. The code is software agnostic, so if you learn some Python, you can use it with commercial or open source software.

The consulting industry is all about worker billable time. If not charging a client in consulting, the person is overhead and costs the company money. Learning a new programming language takes time and this turned out to not be a good fit when it came down to applying billable time to a project. Being new to consulting, there was no real good understanding of how time was charged to a client or overhead to the company. Eventually the question came up, "What have you been working on all this week?" The question was a bit baffling, as all the coding was evidence of the work. The only reply could be, "This AVENUE code."

GIS Recognition: The First Big Win

It was a new approach to use GIS for a client and agency presentation. We went to the agency location and made the presentation. The project manager gave the project history and overview while providing cues to zoom and pan about the site. There was much thanks in the fact that no talking was required by the GIS operator, just navigating the customized ArcView 3.x interface that had been developed. After the presentation was given, there were questions from the regulatory agency. They asked some "what if" questions and as the responses were given, the GIS project would be zoomed or highlighted in a certain way to accompany the response. It was a very interactive session that went on with great inquiry. In the end, a revised action plan was devised and both parties were in agreement; something every meeting with an agency strives for.

When it was all over, the revolutionary approach to use GIS software was a clear home run. It was 1997 and this new software had proven useful in its unique ability to visualize, measure and represent geographic data in such a clear and concise way. It had proven to be an excellent tool for risk assessment, or even at a minimum, just aggregating and presenting GIS layers in a meaningful way. The excitement from the team was somewhat muted during the meeting, but afterwards the excitement could not be contained any longer.

The outcome was less than meaningful to the GIS dude, but when the details were revealed, things made a lot more sense. The summary was that the client did not have to put in an active remediation system. They had to monitor the situation and natural attenuation would take care of the rest. Huh, natural attenuation? The question was asked, "What's natural attenuation?" "Natural attenuation is where the problem goes away over time, naturally." "OH! Oh. Oh..." There was this surprising realization that it would be left in the ground. Surely this had been covered in university coursework, but at this moment it really became a reality.

This would be the beginning of something that would forever take away the shine from the work done with GIS. Technology is neutral and GIS is no exception. It comes down to what it is being used for. There was certainly a feeling of being quite useful and very helpful and this was great. Now the question was transforming into, useful and helpful to who? At what savings to them and with what expense to the environment? Natural attenuation is considered a solution because the risk to the public is low to none, but this means our policy allows giant oil companies to use the Earth as a dumping ground. They make piles of cash and it felt like some of it should clean up the Earth.

On one hand, it was presenting the data as it was. On the other hand it felt like helping the wrong side. At the same time, there was no manipulation of the data. If the data was fully transparent, and it was, who's to doubt the data? Different questions were drifting around in the brain considering this. One provoking thought was that the regulatory agencies were there to enforce the regulations or rules, but what if the rules were too lax? What if they had not caught up with what technology could offer? What if one side was utilizing GIS technology, but the other side was not? Did it matter to have this imbalance? Clearly the answer is yes, given the outcome of this mini-adventure in the career journey so far.

It felt like success because of the way everyone on the team was acting and it is hard to relate to just how excited people were about this outcome. People were ecstatic, as in overly joyful. After the presentation, there was an impromptu celebration at a restaurant. Just the place a young and introverted worker wants to be. This was the first of many victories to come, but this one was really top-shelf and so early in the career. Coming out of university, the goal was to use GIS as an applied technology and the first goal had been achieved.

It was a great time, a time to remember. Like one of the scenes at a Queen's table with food and drink everywhere, everyone laughing and smiling. Everyone was celebrating and reveling in the new technology and praising the new and young talent that made it happen. My boss looked like the winning baseball coach in a world series game and his star pitcher was top tier talent. The client was a large oil company that we all would know if named. They were so impressed with the results that while we were dining and celebrating, they asked what else we could do.

What else could we do? The fact was, for this project GIS was used more as a presentation tool than an analytical problem solving tool. This was accomplished with ArcView 3.x, the lightweight GIS, and we wanted the good stuff. We wanted the heavy hitting software that was learned in university. The new and young GIS wiz said that with ARC/INFO and a spatial analyst license there were more capabilities. The software was agreed to by the client and shrugged it off like no big deal. For a while, it was thought it was just a nice consideration, not something they'd actually buy, was it?

Client Appreciation: The Big Gift

Sure enough, back in the office it wasn't long before a gift arrived. They bought us ARC/INFO and the Spatial Analyst extension as a thank you and hoped for more great things to come. This wasn't a cheap gift at all, some $14,000 for ARC/INFO and $2,500 for Spatial Analyst. This was quite an interesting situation. Software purchased by a client for a successful outcome. One could only think what the savings must have been from the meeting if they could drop $20,000 on software. The number was mentioned at one time, but was probably forgotten out of shame and disgust.

The successful outcome and the expensive software had become a legendary tale about the new kid in the company and GIS. Everyone wanted to be in on it. The hiring boss made several small bound copies of the presentation and filled it with GIS-based maps. It illustrated the project details and became a marketing piece. Oddly, there were no more of these projects to be had. This project was a highlight of the permitting department and GIS was embraced with great regard, seen as a new crown jewel in the environmental consulting industry.

Gifts With Strings Attached

As promised, the oil company purchased $20K worth of commercial software for us. Not sure how this next task was decided, but next we were going to do a site suitability analysis and locate the best place to put a new gas station in North Jersey. This meant setting up ARC/INFO on Windows NT. In retrospect, a client-server configuration was not necessary since there was only one person accessing the data. At the time, the hope was that the technology would grow and expand into the hands of other employees. Did it? We'll never know.

At this point, things were riding on a high, destined for a low, maybe even a double dosing of low. Expensive UNIX hardware and software infrastructure wasn't necessary for ARC/INFO anymore, as Windows NT was a relatively cheap alternative. With this blessing of ARC/INFO on Windows NT came the curse of ARC/INFO on Windows NT. They say never look a gift horse in the mouth and it's true. There were some good learning experiences ahead with ARC/INFO, concepts that would carry over into open source software with ease.

Career Highs And Lows

From this high point, things went downhill in many ways. As life goes, we ride from the highest of highs to the lowest of lows; from difficulty and struggle often comes better things and growth. After 25 years, there's some reflection to support this, but then we also tend to look at things and find evidence to support the claim. Surely, there are plenty of things that are bad and don't have a silver lining, but this story isn't one of them.

We look at events and make the determination if the event is positive or negative. The initial judgment may be cast as negative, and it very well may be, but as time passes and life goes on we realize these things were positive in retrospect. The point is don't let things in life get you down. Enough of the light philosophy, let's get back into the valley.

The Software is Broken?

The first major issue was encountered in this new operating system environment. Using a Windows NT server to provide data to an ARC/INFO Workstation created serious issues between the server and workstation. The workstation would process the data locally, but would also send data back and forth between workstation and server. These were standard networking tasks for UNIX, but on Windows NT there were serious problems maintaining data integrity.

In the case of data network errors, they primarily occurred when the workstation would build topology or clean topology in an edit session. Not always, but oftentimes these network errors would be catastrophic for the data. The integrity of the data was compromised and sometimes odd ASCII characters were inserted into the attribute table. Oftentimes a ? or upside down ? would appear in the database accompanied by some odd ASCII characters. When these showed up, the GIS layer had been corrupted.

When the layer was corrupted, an older version had to be restored. Constantly going back to a layer that was not corrupted, led to the repetitive behavior of backing up GIS files in a numeric sequence. "landuse" would be the current layer, with backups building sequentially: landuse1, landuse2, landuse3, etc. As a little nerd throwback, there is a computer saying and it was first encountered in either Leisure Suit Larry or King's Quest: "Al Lowe says save early, save often.". And wow, did we ever.

After calling the commercial tech support, it was stated, "You should work with data locally on Windows NT because data is not returned from the server to the workstation in the way that the workstation expects to receive the data. This problem doesn't exist with a UNIX server. We [big commercial software company] never fully tested ARC/INFO in a Windows NT before selling it and we have egg on our face from this."

It was hard to believe the software was broken; unfixable due to underlying operating system constraints. It wasn't just hard to believe, it was down right concerning. After seeing the cost of the software bought by the client, it was incomprehensible that this was the current state of affairs.

Centralizing data on a network server allowed for simpler backups and data management as well as providing access to other users. Storing data on a local workstation defeated all of these networking benefits. The question had to be asked, "Where was ARC/INFO on UNIX?", the software that had provided rock solid performance in academia and at the first job at the county. Even being a commercial software fan at the time, it was disturbing and peculiar that such a high-priced and well known software would be sold under those conditions. This was a harbinger of future events, which would ironically, but thankfully, lead back to GIS on a variant of UNIX called Linux. What a great day it would be to return to GIS on Linux.

5.11 When GIS is Viewed as a Threat to Other Disciplines

Previous to the ARC/INFO Workstation server and client network configuration, GIS work was done with ArcView 3.x on a stand alone Windows N/T workstation. Now it was time to use ARC/INFO in a Windows network environment, where the data would live on another Windows server. This configuration meant that the IT Director had to get involved.

By now it should be clear that GIS is software and software uses data. GIS files can be large. Very large. It is this demand for data access that creates a need for robust server and network configurations that can accommodate big data moving back and forth between servers and workstations. Later in the journey, enterprise data distribution solutions are created, but this journey starts with a standard client workstation and data server.

GIS and IT Overlap

It's not uncommon for GIS users to approach GIS software with a technical background that involves knowledge about files, filesystems, networks and internet applications. If GIS users dont have this technical knowledge before working with GIS software, it's very common for GIS users to gain technical knowledge in these areas. Normally these skills, and the knowledge that goes with them, are found in the realm of Information Technology professionals.

How do we say it nicely, but some IT Directors can have a bit of a god complex. In no way meant to be an insult, but a characterization of having exclusive knowledge, and thus power, to enable people with technology or not. They also have an appearance of being omnipresent, by way of knowing who's on the network and what they are doing. This position of power was similar to the GIS niche being carved out by GIS users in different industries.

There are a lot of parallels with IT and GIS, emphasizing just how much GIS is based in technology. Coming from a very technical background, and a Dad who was a network engineer, there was a good amount of computer knowledge known outside of GIS. This double-edged sword was helpful in that it was easier to understand and integrate different GIS technologies. It was, however, undoubtedly seen as a threat to IT professionals because someone knows what is supposed to be exclusive to them. Approach these situations with caution and care.

Having technical knowledge is an automatic threat to most IT people. Don't let it be. This may be obvious to some, but some nerds lack social queues. Highly advisable to bring in donuts to share with them and take them out to lunch occasionally. In the long run, it's important to have this relationship be the friend type rather than the foe type. Only one IT director was encountered on this journey that was really cool with other people having IT knowledge. He is considered an anomaly, but definitely the coolest IT person encountered in the journey. This shout out goes out to you Mr. Wolf, the same person who helped me define GIS better as environmental modeling.

Unpleasant Confrontations

During the stay at the first environmental consulting firm, there was one situation that was rather disturbing. ARC/INFO workstation was going to be run with a client-server configuration. This client-server setup wasn't necessary since there was just one user; it can't be recalled who initiated this setup design. While most of the recollection is spotty, there is a distinct memory of going into the IT Director's office to talk about this configuration.

Once in the IT Director's office, there were sloppy diagrams all over the white boards with arrows pointing to different things. Local Area Networks (LANs) and Wide Area Networks (WANs) were sketched out, connecting each office. The details are not recalled as to what led up to such a violent statement, but there was a lot of tension and no other way to put it but butting heads. It wasn't really clear why there was such confrontation, but the atmosphere was thick with trouble. There was a position of power being challenged by the GIS dude.

The discussion escalated, for who knows what reason, until the peak frustration was expressed by the IT Director. There was a statement made directly that if there was a continued exertion of technical prowess, there would be a broken arm as a consequence. How interesting. It was the first year working as a "professional" and there was already a threat of physical violence. Turns out knowing some computer stuff can make some IT professionals uneasy. Being new to this whole business and employment thing, there was initially shock and then the events were relayed to my supervisor. He was then obligated to tell HR due to the physical nature of the threat. There was apparently a big problem with threatening violence in the workplace. HR wanted to talk about this, and the story was retold. Not sure what came of it other than an apology issued.

It was the first time, but not the last time physical violence was threatened or insinuated. At a later point in the journey, while not an IT Director but a local IT Assistant, there was a coworker who was in a, well, let's call it a group of people who enjoy riding motorcycles together. In a meeting in front of a VP and board member, there was an escalated discussion in which he responded, "If I have a problem with you, I'll just take you outside." He was asked to leave the room, but the statement was a bit alarming considering the source.

Ahh, weren't the good old days just so much fun? Why couldn't we just work together to get something done, rather than be worried about some knowledge? Taking full responsibility, it's possible the GIS dude stirred the pot, but in no way were the resulting threats and insinuations justified. Experience has shown that the people most threatened by outside technical knowledge are the ones who know the least about it. People can be easily upset by the threat of being exposed or being outdone by someone who is not in the IT field.

Build a Good Rapport

These stories are told to convey a point about IT friends vs IT foes. If you are not computer savvy, don't be too discouraged because it may save you a broken arm or being beat up by your local IT gal or guy. Embrace your IT coworkers and if there is any conflict, quickly squash it. IT coworkers can assist you in ways not even imagined. One IT coworker was always helpful and provided old computer hardware that could be recycled to Linux machines to have this open source journey. It may be hard to help IT coworkers, but do what you can. They definitely can help you considering how much GIS has evolved into a technology that touches many aspects of IT. IT coworkers are a group of people you will definately want help from when you need it; and you will need it.

Encroachment is Inevitable

Using GIS is exciting. It's exciting to work through a problem and present the results. Oftentimes, these problems are based in someone else's area of expertise. As exciting as things may get, be careful to proclaim how awesome GIS is because it may unintentionally make someone else feel threatened. It just creates friction that could otherwise be a great joint effort. Software is one thing, but expertise is something a computer can't help with; at least not yet.

With this gift of ARC/INFO on NT there was a chance to use seemingly cool advanced modules found in the thick user manuals in the university lab. Having concentrated studies in hydrology and hydrogeology, there was one module of particular interest called "porous puff". It was a groundwater contamination transport module, and there was a burning desire to try it in the real world; and here we would go. Sort of.

Another lesson learned at this initial consulting firm was not to present any kind of threat or encroachment on someone else's area of expertise. One of the services this environmental firm provided was underground contaminate modeling. At the time, and still today really, there was non-GIS software used to model subsurface transport. With an ARC/INFO extension, users could model subsurface transport from a point source. Without having groundwater modeling expertise, some numbers were plugged in and then some results were obtained. Were the answers correct and accurate? Expertise was needed to get that answer.

The results of "porus puff" were taken to the groundwater modeling expert, who mostly worked alone. As the software and methods were discussed, it was met with indifference. Did the success with the oil company and this new (to the firm) technology create some job insecurity? It's that dreaded job security thing in the consulting industry, where you have to have something valuable to charge to a client or your value to the consulting firm is zero. This time it wasn't the New Jersey "in your face" arm breaking threats, it was another tactic that ignored and starved someone of knowledge in the hopes they fail. Nice. The groundwater modeling effort died right there.

This was another important lesson learned: do nothing to inadvertently intimidate or threaten an existing worker's expertise. Express and have a genuine attitude to work with other people while downplaying the technology. Emphasize how the results are meaningless unless someone with expertise can validate and adjust the model outputs. This also reinforces the idea that the value is in the worker, not the software. Besides, they could just as easily learn to use GIS software and apply their expertise with it, cutting you out.

Slow Down and Enjoy the Journey

Even outside of work, things were taking a turn for the worse. Illustrated as a funny side note, a carpool was shared with family members to the Holmdel, Bell Labs New Jersey area. From there, another car was used to hop over to the consulting firm. The car was my wife's car, a 1989 Ford Mustang, which was affectionately known by her mechanic as *Satan's Chariot*. It had earned this label due to mysterious issues, endless repairs, and the suffering it had put everyone through. Anyways, it's always important to be mindful of our actions, we never know who is in front of us or behind us.

In one hop from work to the Holmdel carpool area, it was Friday after work and there was a rush back to the waiting carpool. Satan's Chariot was in full motion, racing to get back before the carpool was missed. It was decided the car in front was not going fast enough, which regrettably was faster than the allowable speed limit. There was tailgating, swerving around to pass and then swerving back behind the car, acting like a total fool. After all this commotion, the car in front finally slowed down and moved over to the shoulder. While passing the car, there were arms flailing around and a turn to the driver with arms up in the air saying, "What's wrong with you?". If it wasn't clear, acting a fool was 100% accomplished. Thankfully, no rude motions were gestured, who knows what the outcome would have been.

Moving forward and continuing the rushed state, the next thing observed was flashing lights on what appeared to be an undercover police car. The same car that was harassed and attempted to be passed. It turns out the lady was a detective who was not very happy with the behavior she witnessed. What a wakeup call. As the ticket was received, Satan's Chariot pulled away one final time.

While accelerating through the gears, yes, one last time, the clutch broke. The car drifted off to the shoulder and stopped. It was impossible to move the car and the next thing observed was the same unmarked car and detective pulling up behind the car. She walked up to the window and asked what happened. After explaining, she just said, "You're not having a good day, are you?". Maybe it was a lesson that should have been applied in a metaphorical way: slow down and enjoy the journey, don't rush to the destination.

Foolishly Enjoying Exclusivity as the Sole Performer

During this initial first year of full-time employment there was a good amount of naivete and the enjoyment of being an exclusive user of GIS software. There was great enjoyment being the gatekeeper of the data and using cryptic commands in a terminal window to create and edit data. Being able to program in obscure GIS languages like AML and AVENUE also created an exclusive status. This seemed like good job security but it was a foolish mindset.

Concentrating the software around a few users may seem like the right thing to do, but it prevents many people from accessing data that would otherwise be useful to them. Data is data, but people are a wild card. If there is a pile of GIS layers, one person may see and do incredible things while the other may produce something just plain obvious. The pile of GIS layers is the same for both users, but their capabilities are what unlock great things. It's much better practice in an organization to make GIS as inclusive as possible, beyond heavy GIS users.

Thanks to open source GIS software, the cost barrier has been removed and many professionals can work with GIS data in an inclusive way. Web technology was another factor that shifted GIS exclusivity away from operators for basic data viewing and querying. There's no need to make a map of the same layers if people can use web mapping to view these layers against an ever changing study area. This ideology of exclusivity would change later, based on a variety of factors. The final chapter focuses on this area of GIS.

Consulting and the Numbers

Back to the journey and the job, there was a double whammy about to come because not only were there issues with the software, but the job too. With our new Arc/Info on Windows NT and Spatial Analyst software, the oil company wanted to use known station locations and other variables to determine where to place a new gas station. Gravity model documentation suggested using areas that draw a lot of people to them, malls, churches, schools and high-density developed areas. We did practice projects in the university lab finding suitable locations and this was the first real-world application. There were a few areas that came up as "highly suitable" and they were presented to the client. The results were neat but there wasn't any drama. There was no hooting and hollering like the success with the the Risk Based Corrective Action (RBCA) project. There wasn't any fanfare or gifts. Just trouble.

The research took time; unbillable time. In consulting, unbillable time is an eventual death sentence, even given previous excellent performance. Paradoxically, in consulting there is no great performance unless there is billability. People have short memories and corporations even more so. Billable utilization was being constantly calculated and the numbers were being reviewed. A flag was raised about the lack of billable time. What did this mean? How does one get more? What could be done? No memory of the past? Unsure of what this meant, but it didn't sound good at all.

Consulting is tough because it's driven by billable hours, which are the hours an employee can charge to a client. Companies strive for 3X multipliers on what they pay an employee and what they charge in the market. Roughly speaking, one third is employee salary, one third is benefits and one third is profit. Support staff don't control their work but are given a goal to be X% billable per year. If there is no time to charge a project (client) an employee becomes overhead. Instead of making the company money they now cost the company money.

This billable utilization concept was not known at the first consulting firm. This really put a damper on all the fun that was being had, researching and implementing various applications of GIS. Research and exploration are not necessarily billable items, so if you enjoy these things, be sure to find an organization that is a good match.

Time to Move On

Based on this "red flag" news, a new job was immediately sought after and quite astoundingly, one was found on a job board. It was at another environmental consulting firm, but this one was based in Florida. They needed someone to work directly on a client's site. Sounded fun to be saturated in an entire GIS environment. Resume provided, phone call received, flight down to interview with the hiring manager (cool boss to be #4). It couldn't have gone better.

An overview of the Risk Based Corrective Action was given and it really drew a lot of interest. The story about the "successful" client outcome and gift was quite impressive. Who really knows what the big draw was, but an offer was made and it was accepted. It was time to tell the current employer goodbye.

The departure was announced and the direct supervisor who made the hire was completely disappointed. There was a mutual feeling of disappointment, but some joy about the prospects of a new future. Before leaving, the regional director asked to be seen in his office. In an odd turn of events, it was there he tried to get me to stay. This was confusing because earlier the message was that there was not enough billable time being applied. There was the incredible success story that started this whole mess, which apparently had not been forgotten afterall. The regional director asked where the next stop was and then continued to say the new firm was having financial troubles. It was an interesting and unprofessional tactic, but the path was set to Florida.

5.12   Next Hop

In this next hop to Florida, the following 24 years would be filled with vast volumes of entry level work and advanced complex challenges. There was a GIS evolution and an open source revolution about to happen during an explosive time for GIS software. It was finally here that Linux would be discovered along with an amazing suite of open source geospatial tools. It was a time for not only discovering open source software, but also watching the open source revolution rise up and directly compete with the commercial monopoly that had dominated the US landscape. What an awesome time to be alive, some would say.

In the next chapter, we learn about the Big Switcheroo. A big change is coming for commercial software, one that would last for a good 20 years. While commercial software struggled in so many ways, open source GIS projects like QGIS thrived and gained an enormous community of users. QGIS and GRASS GIS would prove to be a powerful combination in the following years, but first we have to get to the next destination in the journey.

6. The Big Switcheroo

6.1 Clickbait?

The Big Switcheroo? Sounds so intriguing or even like clickbait. Rest assured, there is a big change coming and it is the driving force behind the journey to open source GIS software. Like any good story, there has to be a build up to the good stuff. Taking our time to illustrate the painstaking experiences that drove the author away from commercial software into open source software. It wasn't a few petty things or something that was just annoying, no, things got so ugly that there was almost an abandonment of GIS software altogether.

6.2 The End of Commercial GIS Software Use

Environmental Consulting - The Second Company

At this stop, there is finally open source Linux and GIS software. It is where the real fun begins. This next stop would last for about 24 years in a small environmental consulting company as a formal GIS Analyst/Programmer & Manager. Other roles were assumed that ventured into different aspects of GIS software.

The new position offered was a GIS Analyst and Manager, but really the official title in the company was Graphics Illustrator; an old predecessor to GIS positions. GIS was relatively new to businesses and titles were just finding their way into their categorical structures. It took a good 10 years to get GIS titles in a formal status at this next organization.

Settled In and Getting to Work

At this new company, there was an established relationship with a world leading mineral producer in central Florida. The client's operations were heavily dependent upon GIS technology to map land and manage reclamation efforts. They needed consultants—contractors—to work on-site with their staff. At the same time, our firm was mapping land use and jurisdictional areas for land development permits. These weren't small areas; two 20,000 acre sites and a 10,000 site to cite some numbers. There was going to be a lot of work and there was no turning back.

The immediate need was on-site, working with client GIS and permitting staff. The areas this mining company worked in were large, enormous tracts of land that required a lot of GIS effort. There was nothing but GIS work to do and it was awesome. There wasn't a fret about billable time, just lots of work to do. Interestingly, this mining company had a GIS configuration that consisted of a UNIX server and many UNIX client workstations. They were transitioning to a hybrid UNIX server and Windows NT workstation configuration for cost saving measures. This setup proved interesting in that it avoided the network errors found in a pure Windows NT environment.

After experiencing frustrating issues using ARC/INFO Workstation on Windows NT, it was great to be working in a UNIX environment again. Everything worked as advertised and there were no network errors. Even the Windows NT workstations were working without issue when connected to a UNIX server for the data. It would have been neat to know if Linux was substituted for UNIX as the data server, would the NT errors have gone away too? Probably. Too bad it was never tested out. Anyway you slice it, it was nice to be working in a reliable operating system and GIS again.

Pre-development Wetland & Land Use Mapping

There was a lot of work to do and it was the entry level type work. Building data for existing vegetation and land cover was relatively easy. Teams of people with GPS equipment would go in the field and collect GPS data points to delineate wetlands. For more general land use mapping, large 3ft x 3ft maps were plotted at 1"=200' scale for field use and marked up with sharpie pens. Shout goes out to the HP 755CM plotter we used; this thing was an unbelievable work horse. Who knows how many 150 rolls have been put through it but it has to be over a thousand plots. At 1000 rolls, that's 150,000 feet or about 28 miles of paper plotting. Yikes.

Ah, the good old days of low tech paper. For the land use mapping, marking up large field maps was acceptable because the land use generally followed features on the aerial photograph. The data did not need to be collected with sub-meter accuracy, so as long as it was clear on the map what the land use delineations were, the GIS operator could easily map these features. It was a relatively quick and accurate way to get the data built for 20,000 acres of land. These large land use maps would be scanned and georeferenced, and then heads-up digitized. This was a lot of digitizing for the pre-development GIS features, even with efforts to automate line extraction.

Post-development Wetland & Land Use Mapping

The post-development data was an entirely different situation. Part of the post-development data that needed to be created was land use and topographic contours. The contours were created at two foot intervals, with some one foot contours made to add design detail to wetland features. To say there was a lot of digitizing for this task is an understatement.

Anyone remember that cool owl in a tree with a lick-able lollipop, and his "How many licks does it take to get to the center of a Tootsie Roll Pop?" question? Well, just how many clicks does it take to make one foot topography for 20,000 acres of land? One, two, three million? Who knows for sure, but it would be interesting if our wrists had a click counter. Be careful not to damage your wrists during these early days of mass data entry, the injuries are real.

There was this one time that an emergency called for redoing an unbelievable amount of contours to accommodate some regulation or request. The words still echo through the phone to my boss, "This isn't possible even if we stayed up all night and clicked till morning." There were so many hours put into digitizing data that when the eyes were closed, flashing lines and star-like speckled points would dazzle the brain. It can get pretty intense. Beware of massive entry-level GIS work efforts and be sure to have a good ergonomic setup.

In the GIS Grind

It was the early 2000s, and as time went by, more time was spent working in the office than on the client site. We were mapping 20,000 acres of land and work was in high production mode. In the office we had a Windows NT client and Windows NT server setup. You know what this means, frequent ARC/INFO bombouts and back to "save early, save often". There were many backups made of the working files, sometimes corrupted data set us back. It was the same networking problems from the past, but there was word of a new product coming out - "ArcMap".

ARC/INFO Workstation data corruption was no small issue. It wasn't just an inconvenient bug or something that was an annoyance. The realization of losing data integrity on 20,000 acres of polygon data was a gut wrenching moment. We were not mapping 10 acre parcels with four land use polygons that could be quickly figured out if something went wrong, these were the biggest land mapping projects to date. Rolling back edits to a previous state is like digging a hole four feet down only to fill it in and dig the hole again.

There was so much work to do that we needed to hire another GIS worker. Together, the two of us did the work of three, maybe four people. We worked 50-60 hours a week regularly, with one timesheet logging over 80 hours. It was an unbelievable time that called for an unbelievable effort and dedication. We had a great team consisting of office workers and field workers. The workflow was practically flawless and we rocked the job as a small group of dedicated, tight-knit professionals. Seems like these days are gone and absorbed in giant, faceless corporations, churning out work. These were considered the high production days in this journey, with enormous projects, intense amounts of work and major accomplishments.

The Odd State of Commercial GIS Software

During the GIS grind, there was an odd configuration of two pieces of commercial GIS software. This is of special interest for the upcoming Big Switcheroo. The combination of ARC/INFO Workstation and ArcView 3.x was very prevalent. We used ARC/INFO ArcPlot to automate labeling large plots for field use and ARC/INFO ArcEdit to manage polygon layers with topology. For large scale production efforts involving massive editing sessions and automatically cranking out large map plots, ArcEdit and ArcPlot were great.

When it came to GIS with a GUI, ArcView 3.x was being used heavily used in the US. It lacked auto-labeling like ArcPlot, but it had a pretty good user interface to get a map out. Ease of use with shapefiles gave it high user counts, but since shapefiles lacked topology, data from this software was always questionable. The fact that there were two software packages sold to users - Workstation and ArcView - was another peculiar situation created by the commercial GIS software company.

In this era, ArcView was known as ArcView 3.x because this incarnation stopped at version 3.x and it was reborn in a new programming language and interface called ArcView 8.0; yes not 1.0, 8.0, in order to fit existing product line labels. ArcView 3.x was the last version in the AVENUE programming language before ARC/INFO and ArcView were merged into one software called ArcMap. This software had different tiers that unlocked additional GIS functions (funny - ha-ha funny).

6.3   First Encounter With Open Source Software

Finally! Without much of a lead-in and seemingly out of place and out of nowhere, Linux was stumbled across. There wasn't a definitive starting point with Linux and open source software that can be recalled, although credit is likely due to a friend met while working on the client site. There was some intrigue into this open source software as an alternative to the Windows operating system.

It was around the year 2000 where Linux and open source GIS were explored. There was GRASS GIS that had an appearance like the commercial software ARC/INFO, but had an amazing list of features and functions. It was heavily based in the raster side of GIS, while the commercial GIS software had its strength in vector data. There was also GDAL/OGR software, but it was only briefly explored during the initial investigation of open source GIS on Linux. This open source software looked cool and very technical; it would be explored and heavily used in the future. There was a GUI driven Quantum GIS that was pretty cool and later it would be known as QGIS. It was in its infancy at the time, but what a gem it would turn out to be.

Some old hardware was recycled at home and Linux was installed for exploration. It was cool alright. The same as UNIX, which makes sense since it is a UNIX clone. There was some home, hobby usage of Linux to learn about web serving and other geospatial software. It was pretty powerful to be able to set up a Linux machine and serve some GIS maps in a webpage. At this time, web maps were a relatively new technology as the dot com bubble popped not that long ago. The web mapping would later be sold as a service for a team to collaborate, but we're getting ahead too quickly. Just around the corner, explosive growth in GIS is coming.

6.4 GIS Software Goes Mainstream

Here we continue to take a small pause from the GIS journey to discuss the massive changes happening at the turn of the 21st century. There's a lot more going on than just a commercial software rewrite. We notice that from a US perspective, GIS users really start to grow in both demand and supply. GIS started becoming very popular, just like the first supervisor suggested at the quasi-public utility company. He was right in that there would be rapid uptake of this technology in many industries, but the reason wasn't really expressed in detail. Some of the reason is certainly due to the fact that there was a technology bubble at the end of the 1990s, known as the Dotcom Bubble. This bubble exposed a lot of technology to the public and industry. It may have helped fuel GIS exposure and demand, but let's take a look at some other strong influences.

Commercial GIS Transitions to Windows

This commercial GIS company's move from UNIX to Windows NT was a clear signal that they were chasing the largest market share of desktop users. At this point, Windows was on 95% of the desktops, a cash-incentive statistic far too juicy for a commercial GIS company to ignore. ArcView 3.x was lower cost than ARC/INFO and didn't have the networking issue ARC/INFO had with coverages. As this commercial GIS software company shoved the UNIX OS and the users to one side, we could see the funnel of cash flow being collected by Windows and put right in their bank account. The fact that their flagship product, ARC/INFO workstation, was now available on a low cost Windows operating system helped accelerate their market share.

Interestingly, Apple's Mac computers were known for their robustness to work with large audio and video files, but Macs were also given the cold shoulder just like the UNIX base. Why write code for all these undoubtedly more robust and reliable operating systems, when there was an existing monopoly sitting on 95% of desktops? It made no difference that Windows was the worst operating system of the three, what mattered was which one could be used to sell the most GIS software. Quality selection of something as vital as an operating system is not an option when the goal is to sell as many copies of software as possible. As this goal was chased, the quality of the software fell off a cliff with the next software rewrite called ArcMap.

Three Problems for Commercial GIS

As ARC/INFO and ArcView 3.x gained popularity on Windows, and GIS popularity grew exponentially, it was clear this old software ported to Windows needed to be re-written. Three problems with ARC/INFO Workstation needed to be addressed if this company wanted to maintain and enforce the GIS monopoly they had created in the US.

The first problem was the broken network functionality of ARC/INFO Workstation in a Windows NT environment. The second problem was the lack of a, now standard, pretty GUI for ARC/INFO Workstation users. The third problem was that ARC/INFO Workstation (C+ and FORTRAN languages if recalled correctly) and ArcView (their soon-to-be dead Avenue language) were written in two different languages. This meant they could not be functionally integrated. A rewrite of the software was necessary to resolve these three issues.

Gimmick Software Extensions

At the turn of the century, ArcView 3.x was in widespread use. Market saturation had created an opportunity for the commercial GIS software company to create "extensions" to sell to different industries. ArcView had started in 1991 and by 2001 this software was everywhere. The commercial company was busy putting annual user maintenance monies into marketing campaigns that convinced industries to buy various GIS extensions for specific work types. If you thought the base software was expensive, wait till you load up on some extensions.

Why is it that many commercial GIS software extensions are an "analyst"? Aren't analysts people? These modules are sold separately, such as: spatial analyst, 3D analyst, network analyst, geostatistical analyst, tracking analyst and business analyst. Don't forget your productivity extensions, like publisher, data interoperability, data reviewer and workflow manager. A data interoperability extension? In open source we call these open data formats.

It's just obnoxious. The whole setup. Expensive software and annual maintenance. Tiered software with different costs for functions, extensions that sound useful but are just bundled GIS functions with buzzwords. There is a place for extensions, but the sheer number of them coupled with the expensive license tiers can make the whole thing quite distasteful.

6.5   Incoming Surge of GIS Users

Maybe more interesting than the software saturation is the user saturation. In the future, this GIS user saturation would have to be addressed and formatted for industry standards. At the turn of the 21st century, GIS software was continuing to grow in industry and educational institutions, which meant there would be a lot more GIS users in the wild. There were mostly three titles used in the industry: GIS Technician, GIS Specialist and GIS Analyst.

Interviewing Users From the Wild

Since part of the current job duty in the journey was GIS management, a GIS hands-on test was created for the interview process. The test consisted of a basic task to digitize some topologically correct, adjacent wetlands polygons in a specific coordinate system. Then the candidate had to create a map. When interviewed, candidates were all over the map (uggh, a map pun?) with this task.

Some could perform the test with ease, while others would struggle and walk out. Yet others would struggle and persist until they figured it out. There was always a strong preference for those that asked questions, struggled and persisted. Overtime, the test became a barometer for how candidates would perform under pressure. If candidates could dig in under stress, ask questions and persist to the end with some help, then they had a good chance to get hired.

During this process of interviewing, the GIS field was in full bloom. It was somewhat chaotic as labels and titles were flung around carelessly. One of the outputs of the chaos was that anyone who could press buttons in ArcView called themselves an analyst. As a manager, there were all kinds of candidate encounters, but one is quite memorable. It can't be recalled exactly where in the story, or under what conditions, but it was definitely during the inundation of GIS users into the field.

Self-Titled Analysts

We were looking to add another GIS Analyst and an interview was scheduled. Browsing many candidate's resumes, it was clear there was plenty of self-titling occuring in the GIS field. The field was saturated but industry had not caught up. In came a candidate with a resume full of GIS jargon and a GIS Analyst title. Trying to dig into the GIS Analyst title, the candidate was asked for a simple example illustrating how GIS was used in an analytical capacity. Overlays, summaries, buffers, any of the GIS operators with a problem to solve would be just fine. The reply was stunning but driving the point home about self-titled "Analysts".

The analytical example cited was, "I ran AML scripts that converted coverages to shapefiles." Well, this was something, potentially. This wasn't really an analytical application of GIS software, it was just some scripting that ran conversion commands. Which, okay, it's still code to solve a problem. Let's dig further, "Oh, did you write the script? What can you tell me about it?" to which the reply was, "No, I just ran it." Just ran the script? That's the analytical highlight? Unbelievable.

Ah yes, the elusive GIS Analyst running conversion scripts; hadn't come across one of these in the wild yet, but sure enough, it was right there. After this rare encounter, and it wasn't the first, it was clear the GIS field was overwhelmed with users. Maybe this person thought they were an analyst or was assigned the title by an unassuming manager. Either way, now what do we do? What could possibly fix this mess? It wasn't just one incident, "GIS Analysts" had come to be anyone who could run the software. Bummer.

The Need To Differentiate GIS Software Users

There were lots of people entering the GIS field and using the software now. It became clear that there needed to be a way to differentiate GIS users from GIS experts, or practitioners from professionals. There are people who use GIS software as a tool for their job much like a word processor, spreadsheet, database or presentation software. These users of GIS software are just that, users, not concerned with software or systems, just access to the data to do their jobs. So in this way, we are trying to identify one group of users, being GIS software users who use GIS as a tool for shallow or narrow applications.

On the other hand, we have GIS software experts, who understand GIS software and the data that goes with it. GIS experts work with advanced GIS applications that go deeper than surficial use. Sometimes advanced GIS users find themselves encroaching on other areas of expertise, just remember the warning and be inclusive. An example is someone who can master the hydrologic tools in GIS modeling, who may or may not be a hydrologist by profession. Within this field there can be experts in different areas, hydrology, forestry and remote sensing to name a few.

GIS professionals in the context of this book will be left undefined as anything more than people who use GIS software as their main profession. The key to identify a good GIS power user is to look at what the GIS user has done with the software and how it has been applied. See what problems have been thought through and solved with geospatial technology.

Industry had not caught up with the flood of new users by way of formal titles and responsibilties. Users ranged from basic to advanced, casual to professional, all using titles and labels that the industry was trying to apply on the fly. The GIS Analyst who ran someone else's script as an analytical career highlight is one of the labels looking to be adjusted. Certainly there were organizations ahead of this curve, but in general it was a flood of people all claiming GIS experience. The claims are fine, maybe, but they do need to accurately match industry expectations.

What Is a GIS Professional?

Among this GIS user inundation there was one question to people, "What is a GIS Professional? Is there a Geospatial Professional? An organization came into the GIS field called the GIS Certification Institute, which aimed to distinguish GIS professionals from GIS practitioners.[20] This is the main user division we talked about, people who use GIS as their primary field of expertise, compared to people who use GIS as a tool for their primary area of expertise. It has become necessary to make this distinction in user types, based upon experience as a user in the community as well as a hiring manager.

Is there a GIS profession? Sure, it's a paid occupation, but we could also argue it's just software. Is Excel a profession? If a person is using GIS software on a daily basis and focused in a field of science, are they a GIS professional

---

[20]GIS Certification Institute https://www.gisci.org

or are they a scientist? If an engineer is using GIS data as part of floodplain
modeling, are they a GIS professional, or are they a water resource engineer?
What about the person administering a GIS-based web system? How about
a GIS manager that doesn't do much GIS work, but interfaces with GIS staff
and organizational needs, are they a GIS professional? What about the GIS
Analyst who has an education in GIS concepts and uses the software as their
exclusive job function? Some answers may be more obvious than others, but
there are definitely gray areas.

6.6   GIS Certification Institute

History

How do we deal with this enormous GIS user saturation? Great question. GIS
certification efforts started in the late 1990s, but by 2004 the GIS Certification
Institute (GISCI) was a functional organization. A goal was to provide GIS
software users with 3rd-party validation, separating the user chaos into two
groups. The process went through many challenges, but emerged with a way
to evaluate users of GIS technology. If certain criteria was met, users could
wear the *GIS Professional - GISP* badge. At this time it is $100 application fee
and $250 testing fee. GIS certificate renewal is $95 annually. Unless there have
been some changes, this is only for US citizens.

GISCI did a good job partitioning users into two groups: professional or practitioner. This was not an easy task to do. The history is illustrated on their
website and shows the curious challenges that come with defining the GIS
profession. Even trying to qualify what a GIS professional is in a book, chapter, paragraph, or even a sentence is why we won't define it. We don't have to.
There are gray areas for GIS user classifications and that's okay.

The GIS Certification Institute did a good job creating a validation process that
could help users prove their professionalism. It also helped hiring managers
know if a person's experience had been reviewed by a governing body as well,
validating user claims a little more. This was a good start to help define both
the supply and demand side of GIS operators, but how about the bigger and
broader picture of GIS? What was happening with the technology in industries, government or even with regards to regulations?

Interestingly, there was a Geospatial Data Act being discussed that had potential regulatory implications for people using GIS software. GISCI had an
active presence on behalf of the software users. Both the GIS community and
the GISCI had at least three complaints with the Geospatial Data Act: 1) the
bill lacks clear intent 2) possible reduction of funding for GIS data from gov-

ernment (section 10) and 3) language potentially requiring surveying certifications for federally funded geospatial projects. The last one has a potential to reduce available work for GIS Professionals by requiring a professional surveying license to conduct certain types of mapping work.

The GISP Application

Not being a fan of labels, it wasn't until 2006 when this designation was applied for. In the early days, there were two ways to apply: under the formal application process or a grandfather entrance. While eligible for the grandfather entrance, the application was filled out under the formal process with the mindset of earning it rather than a gift through time.

The first requirement for the formal application was to have three (now four?) years of active work using GIS software. In addition to a minimum length of time using the software, points had to be earned in several different areas. The three different areas were designed to promote a well-rounded GIS Professional.

The first area considered formal GIS education. This required college or university credits in GIS subject matter. Courses needed to be provided to GISCI to substantiate the claim. For renewal applications, since formal education can only be considered once, this came to include conference attendance, online learning, or ways in which the candidate can receive information and grow their knowledge base.

The second area was orientated around contributing information to the profession. This requirement was to help people engage and share information with the community. Actions include publishing maps, writing articles, presenting GIS topics at events or working with the open source community are ways to contribute to the profession.

The last criteria was professional experience. This criteria looked at two components relating to professional experience. The first examined what the person was doing with GIS. Was the user constantly making maps or was the user engaged with the software at a deeper, analytical level? The second component looked at how much hands-on experience the candidate had per week at the given GIS level. For example, 20% of time spent with GIS analysis carried more weight than 20% of time spent in cartography.

Overall, this was a good system to separate two user classes; the practitioner and the professional. The organization was also helpful, by way of its certification design, to encourage the sharing of information; an essential component for a profession to really thrive. The GISCI is successful with its GISP mission, as evidenced by organizations recognizing the credential and requesting that applicants have this credential for a GIS or GIS related position. The credential has also found its way into requests for proposals (RFPs), asking GIS workers have a GISP certification.

GISP: A Desirable Certification

As time went on, this GISP designation was something that people were talking about and were interested in obtaining. Some people were using GIS on a daily basis and other people were using it secondary to their primary job duty or expertise. It was also apparent that bids for GIS work requested this designation and job positions had a strong preference towards candidates with this badge. It was being recognized and sought after.

The general sense was that non-GIS users 1) found GIS to be fun and 2) the GISP credential was recognized and requested by other organizations. For those who enjoy a long chain of initials after their name, GISP was a good one to have. It became a fashion symbol to wear after one's name, almost stating openly that once you had the badge, you had the cred. For some yes, others, not at all.

In people's quest for initials longer than their name, GISP became a sought after credential. GIS was now the cool kid on the block and everyone wanted to be involved with it. People were applying for the credential who should otherwise be ineligible. This may sound petty or like some kind of resentment for many people obtaining the same certification initials, but the reality is that it got embarrassing to wear these initials. Some of the badge wearers were questionable. There was no desire to be considered similar to the rest of mainstream non-professionals now flying the GISP flag. Data was always handed to them in a proper format, so they didnt know how to get their hands dirty and figure out what what's wrong with the data.

There seemed to be people everywhere gaining the GISP certifications, but some didn't seem to be the intended recipients of the designation. One example was an engineer working as a project manager on floodplain mapping projects. This person was mostly managing contracts, but did some GIS tasks also. The person's primary job function was a surface water engineer. When this person needed to plot lat/long coordinates for a project, they were baffled about why coordinates were in the Gulf rather than on land. The worst part wasn't that a simple error was made, but that there was no ability to figure out what went wrong and how to fix it. Attempts at resolving the issue illustrated how novice the person was with GIS, yet they would soon have their very own GISP badge to wear.

Mixed Feelings on GISP Certification

Skipping ahead in time for a moment to 2016, the pool of certified GISPs got larger and it seemed anyone could wear the badge for $150. Who knows for sure, but it seemed that some of the issue was validating what candidates said in their application. Somewhere around this time, 2016, an exam was introduced as part of the credential. There were also changes to the certification process at this time too. The length of time the certification was valid for was almost cut in half and the renewal fee for certification about doubled. Double the cost, almost halving the duration. That's some good money. People were starting to wonder what value of the badge was for the cost. Was there a newsletter, industry updates or salary gauges? Maybe, but it seemed the information had to be sought after.

There was no test initially in the 2000s, but over time this would be a cornerstone. The test that got implemented in the process proved problematic and covered odd topics that were related to geospatial infrastructure. Note that geospatial is larger than GIS, yet the certification was for GIS professional, not geospatial. The test always seemed to be going under a formal review for modifications, but the test had never been personally taken. A colleague and GIS user with over 15 years experience took the exam and didn't pass.

It was odd because applicants were required to go to official testing centers, similar to a Linux or other computer certification testing center. At the time for the colleague, there was no study guide available. Does this occur in other professions? Do other professional certification exams exist without an accompanying study guide? There were sample test questions available at the time and the author only got three of the five correct. These review questions were the only material provided to study for the exam. The test was reissued at a later point in time and questions went out to the community for help.

Tests are tricky because we need a way to show or validate someone's knowledge, but tests may not be the best way for all takers to show this. Tests seem tough to develop for the geospatial world. GIS is an applied technology with many layers from data, to desktop, to database servers and cloud-based web servers. If someone uses software as a problem solving tool, how can their skills be tested without providing the software and a problem to solve? Yes, there are some underlying basics of the profession, but beyond that it's how the software is used.

Then we're back to the abstract question: what constitutes a professional? Is it experience and knowledge or is it exclusive work in the GIS field or software? How should it be qualified or quantified? Do we just test basic knowledge that should be learned from a book, things like projections and topology? That seems not quite right. What about cartography and data usage versus data creating and modeling? Should there be a distinction even between cartographers and modeling or analytical professionals? How does one test cover both aspects? Tests are tricky for GIS technology because ultimately it's a software test that isn't about pushing the right buttons.

The early GISP application process in the 2000s seemed better at identifying professionals without the test. However, at the same time we had engineers wearing a GISP badge who did not much more than casual GIS use and manage budgets. The initial process seemed fair, but needed investigations and affidavit-like verification statements from supervisors, ensuring applications are accurately representing themselves. Even this idea is a slippery slope because management has a motive to get the certification as they, the organization, benefit from the badge too. Having a GISP employee opened the door to bidding on GIS projects. There's the supervisor who may just be ignorant to the whole GIS thing and sign off on an employee who uses GIS casually. None of this is perfect, but maybe there should be a community validation or anonymous alert for fake GISPs. But then we could end up with factions and people helping people get certifications. There doesn't seem to be a good answer.

All of this, and for what? To root out a few bad actors and give some people self-worth by way of a four letter badge? The badge is only important because hiring organizations perceive it has value. Overall, the GISP organization was a necessary entity in the GIS profession because the field grew so wildly. The organization succeeded in adding some order to, if nothing else, raise awareness about the two types of users. Some GIS analysts and power users do not give much validity to the GISP certification and that's okay, because GISCI did do a good thing in the bigger picture. The only way to really cast a judgment is to get involved, so at least get involved before you form your own opinion.

6.7 Farewell Commercial Software

The pivotal point in the journey was here. This big user boom was being partitioned by the GISCI, while the commercial software company was rewriting their software for the next generation of users. This would bring a whole new chain of events that ultimately led to taking the fork in the road to open source GIS software. A chain of events that would either drive the author out of the profession, or move to better GIS software. It was time to say good-bye to commercial GIS software.

Leveling Up to New GIS Applications

Despite the network errors on Windows NT, there was still a naive admiration (not quite the right word) and a fair amount of advocating for this commercial GIS software. The journey continues as we are still mapping a 20,000 acre tract of land with another similarly sized project for wetland delineations using AR-C/INFO. These were massive efforts handled by a tight team of professionals who, throughout the process, laughed and had a lot of fun together.

As work in the projects deepen, other aspects of the industry become apparent. On these large projects, drainage area delineations were performed by engineering firms. The reason watersheds needed to be delineated was to provide assurance that post-development conditions matched pre-development conditions; meaning the land had to generally be put back the way it was topographically. For example, the streams and basins had to have the same Strahler ordering pre and post development.[21]

Once it was realized that watershed delineations were a critical part of the client's regulatory compliance, the excitement was uncontainable. The prospect to model drainage areas and streams was just too much to consider. Since the university days, now about five years in the past, there had been an overwhelming desire to use GIS as an applied technology to surface water modeling. The nerd buried in the thick manuals in university was now drooling at the prospect of applying this technology to the real world. Finally, the dream to combine hydrology studies with GIS was about to come true. Well, not quite yet with commercial software.

Dead End Hydrologic Modeling

Hydrologic modeling, or more specifically drainage areas, were typically done by engineers with related software. The results were more about linked basins than actual drainage network modeling by topography. There were drainage area outputs, but the corresponding lines to link the basins didn't follow topography or make sense from a flow accumulation point. This was discovered later and was initially very confusing from a natural drainage perspective.

With great ambition and excitement, this new pool of opportunity was lept and dove into with unbelievable enthusiasm and excitement. The problem was, there was no water, just cement eight feet down. It was just like in the old cartoons when a pool drained right before the toon dove right in. Given

---

[21] Learn to apply Strahler orders, build watersheds and more, read *QGIS for Hydrological Applications* https://locatepress.com/book/hyd2

this was 20 years ago, the details of this part are sketchy.

The best available data at the time was two foot contour data and it was loaded into ARC/INFO. Modules were run and results were obtained. The outputs looked okay with some obvious errors that were unfixable. The results of the drainage delineation were presented to the client, and much like the suitability analysis to find a new gas station location, the reception was not so great. Not sure why, it was never explained. So much for living the dream and all that stuff.

There was one step in the process that was peculiar, and it was to fill in all sinks in the DEM. This little step would be thoroughly reviewed in the following years. We would have to revisit hydrologic modeling in the future with far superior software. What a bummer. The dream died quickly but the embers were still smoldering, rekindled later with open source GIS software.

Shocking Hydrologic Modeling Discoveries

The first problem was understanding how the water was being routed down the terrain with the commercial software. The overland flow module in AR-C/INFO was based on an algorithm from the late 1980s. It was not very intelligent and it only routed down the steepest and shortest route, known as *Single Flow Direction (SFD)*. With SFD modeling, when water flows off of a small hill, there is no sheet flow effect. The output would be a single line of cells.

In *Flow Accumulation* modeling, the flow becomes concentrated in lower elevations because of the terrain characteristics, not because an algorithm clumps cells together. *Multiple Flow Direction (MFD)* modeling is a great tool to illustrate sheetflow because MFD models sheetflow very well as it converges into linear flow. MFD is more advanced and much more useful in overland flow modeling.

When this difference was realized years later, MFD vs. SFD, a call was placed to the commercial GIS technical support asking "What's up? Why no MFD?". The call was escalated to a higher level, to what was believed to be an authority on Spatial Analyst. He was in charge of the raster support staff. He went on to say, and it wasn't the first for this commercial software company, "we can't do that, it will break the software". What kind of spaghetti code was this ArcMap?

Later, around 2020, it was noted that the commercial GIS company finally did implement MFD. Wow, congratulations on getting into the 21st century 20 years too late. It wasn't the first time commercial GIS software lagged behind open source and was late to implement necessary features. How many projects suffered with poor modeling results because this commercial company was afraid to "break their software"? How many people believe this commercial software is really best-in-class? Far too many.

Single flow direction (SFD) was bad enough, but the real deal killer was the simplistic approach the flow accumulation model used. The algorithm could not route water out of a sink or closed depression in the elevation model. A sink, or a single cell is called a pit, is a value in the DEM that has a lower value than all of its surrounding cell values. An example would be a bowl-shaped wetland that holds some water but after a while the water fills and exits the wetlands depression.

Florida has some unique terrain because most of the landscape was formed under water with only geologically recent overland flow to shape the terrain. As a result of this atypical land formation, Florida is full of flat terrain with many depressional features. As research continued, some light was being shed on why there were so many difficulties getting good results with commercial GIS software. Their solution was to fill all the depressions in the DEM so their algorithm would work. In other words, because the flow routing SFD algorithm is so simplistic users must alter their survey data to get the software to work. Not cool.

All this history to get to the pivotal point in the journey. There were plenty of problems up until this point with commercial software, but none were severe enough to want to jump ship entirely. This was about to change with the upcoming Big Switcheroo.

6.8   The Big Switcheroo: Take It or Leave It They Said

The New Commercial GIS Software Revealed

Remember the three problems the commercial GIS company had to solve as GIS user growth accelerated? They were: 1) broken ARC/INFO on Windows NT network, 2) no pretty GUI for flagship product ARC/INFO, and 3) two similar products written in different languages (ARC/INFO & ArcView 3.x). So around the year 2000, they rewrote the software, relying heavily on the Windows subsystem, and started asking users to beta test their software. A devious way to get free product testing and feedback because some users wear this beta badge with great pride and boastfulness.

This marvelous product rewrite is called ArcMap. A 32-bit quirky desktop GIS that had a 20 year run before being replaced with ArcMap Pro. The first peculiar thing about this brand new product release was the version number. It was launched in late 1999 as a brand new software rewrite, so it makes sense to start the version number at 8.0. Huh? This goes against every software versioning philosophy (except maybe Windows) that aims to help the user know how stable and mature a product is.

When senior staff at the commercial company were asked why a new software release started at version 8.0, the answer was to fit in with their existing product line number. What? Fit the product line?

This would be like starting an interstate mile marker at 8 rather than 1 (okay zero). Why not pick me up at 8:00? Rather than 1:00? If we got one dollar an hour vs. eight, what's really the difference? What's the point of numbers anyway? One or eight, their just numbers, who cares if they're supposed to be meaningful, especially in software development.

As far as commercial GIS is concerned, there were product lines to match up with and that's far more important than indicating if software is in the fledgling or mature state. Of course, saying it needs to fit in with the existing product lines is just a nice way to say it's cleaner marketing.

This was an outright marketing ploy, set to deceive users by inferring some kind of software stability. It should have been advertised as "Version 8 - It's Just a Number". Even a non-developer knows that software versions are designed to indicate software development states. This is really an incredible thing if we pause and think about it. It's no mistake, it's not an oops. There is a *product line* to make cohesive and sell as a collection of software; we know it as vendor lock-in.

Not only had ArcMap been rewritten in a new language, but it also claimed to consolidate the ARC/INFO workstation and ArcView 3.x into one product. Ah, but the catch was, there was an Arcview (now called Basic) at $2,500, Editor version at $7,000 and an advanced license at $14,000. The editor and advanced licenses could work with a spatially enabled Microsoft access database, later to be deemed terrible by everyone and a new format file geodatabase would emerge. This spatially enabled Microsoft Access database would have topology, one of the essential things in GIS data that shapefiles lacked.

Interestingly, this commercial company kept on using the archaic shapefile for the ArcView (Basic) version. In the end users got one product interface, but their GIS functions were limited based on which tier software they purchased—Basic, Editor, or Advanced. What was the difference between these tiers? Money. Access to geospatial functions was dependent on how much money an organization had to spend.

Insult to Injury - Throw Out Your Duplicate Licenses

Earlier we took special note that it was commonplace for an organization to purchase two pieces of software: ArcView and ARC/INFO Workstation. It was now time to consolidate these two products, but what did that mean for customers?

So here we are with organizations that have bought two pieces of GIS software for one user. Think about that for a moment! One software package, ARC/INFO Workstation manages topologically correct data and is useful for editing vector data, but this software requires scripting and is very cumbersome for daily cartography use. The other software, ArcView 3.x is GUI driven and much more user friendly, especially for maps. One user needed both pieces of software and each software had its own annual maintenance and specific extensions.

For example, they sold a Spatial Analyst extension for ARC/INFO and one for ArcView 3.x. There's the initial purchase and the annual maintenance. It's the annual maintenance that is the real money drain over time. If the maintenance is not paid annually and expires, the software is not eligible for upgrade and users lose their customer support. So many organizations, so many annual fees to collect.

These two software pieces were kept in parallel at most organizations that used commercial software. It's not really clear why this was considered acceptable, other than users being held hostage by vendor lock-in in the US. If we think about it for a moment, why would a commercial GIS company have two separate softwares that perform the same GIS functions? One is much cheaper than the other one, but the real problem seems to be coming from the UNIX world and mashing everything into less stable Windows environment as hurriedly as possible.

This two-software approach seems like a function of poor planning and evolution; or was it? If the $14,000 software did cartography just fine with a nice GUI, why buy and maintain another software and all its extensions? Another way to look at it is extracting more money from users between two software products that should really be one.

Now for the solution and problem. The solution was one, single ArcMap software that had three tiers: ArcView, Editor and Advanced. The ArcView 3.x mapped to a ArcMap ArcView (now basic) license and ARC/INFO mapped to the ArcMap Advanced license. ArcMap, ArcView, and ArcMap Advanced were the same software when you fired it up, you just got access to different tools based on what level you bought ($2,500, $7,000 or $14,000). If a user has an ArcMap Advanced license, then there is no need for an ArcMap Advanced license because ArcMap Advanced has everything in ArcMap ArcView. There's no need to have both licenses, but both licenses had been purchased and were still being maintained. It's one software application that launches, providing "tiers" of functionality. Make sense?

Then one day the commercial GIS software company says this new software is what you have to use going forward, and by the way, none of your scripts work anymore. Your existing licenses will transfer into their new product line. The pricing would stay the same, and remember, these prices came from the UNIX software environment. Now what of these unnecessary licenses?

Now that two products were combined into one, there was no need for the ArcView license anymore. It was now considered extraneous. This "left over" license issue was a bit much to take; they reorganized their software and we're left with extra pieces. Regional representatives were contacted to see what could be done. When the commercial representative was asked how to turn it back in, or apply some credit for no longer needing a software they consolidated, the request was laughed at. They said to give it to someone else in the organization. Curious how the solution was now part of an expense to the organization by way of annual maintenance cost?

Asking for some consideration, anything, for this now wasted money seemed reasonable. Instead of listening to a customer's request, it was met with complete disregard as they commenced to laugh at their customer. This seemed quite rude given all the struggles and patience customers have had with this monopolistic giant. They had taken so much money from users, was there nothing that could be done to help offset this consolidation?

If we wanted to spend money and buy extra licenses to pass around, we would have. This was an unbelievable move that kept a lot of licenses under maintenance payment that companies otherwise weren't planning on using. Think of the 350,000 customers they had and how many extra licenses were paid in "maintenance". Many people felt that the whole thing seemed too rooted in making money and less about quality software for the profession.

Adopt It, or Else. Or Else What?

The time had come. It was time to adopt ArcMap. First support for UNIX ended, then support for coverages and finally ARC/INFO Workstation came to an end too. The nail in this coffin was a bit depressing, given all the learning and fun. ARC/INFO was put into maintenance mode, which meant only major bug fixes; but not that pesky network issue? Of course not, it can't be fixed after all.

There was no more avoiding it, we had to move to ArcMap and the new *personal geodatabase*. Personal geodatabase, what is this - a geodatabase only the person can use? We can have personal matters, but the personal geodatabase was named this to contrast the now best selling buzzword, *Enterprise GIS*. The personal geodatabase was later abandoned for *file geodatabase* because it turns out cramming GIS data in a Microsoft Access database is a terrible idea.

ArcMap 8.0, the first release, was so severely flawed, buggy, and unreliable that our organization could not use the software for production use. There were simple things like selected features that wouldn't unselect. Then there were the application crashes. Lots of them. It didn't matter if users were editing data or not, overall the software was very unstable. It couldn't be relied upon to meet customer demands and deadlines. Competitors in the marketplace tried to capitalize on frustrated users who felt stranded.

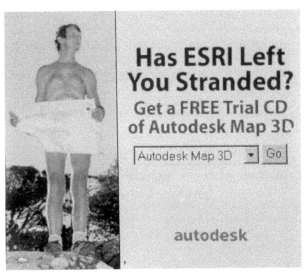

Figure 6.1: Seize the Moment. A competitor's ad from the time, taking aim at the perceived vulnerability.

The ArcMap for Windows rewrite was a shiny new object for GIS users. It was a refresh of what was rather clunky software. This new software could reach a much larger audience as this facial overhaul fixed a lot of the previous image issues. The feedback loop would continue with the next generation of software, but over time, true quality would catch up with this commercial software. A long, slow and solid revolution was occurring and a huge fan base and support would push open source QGIS to be a main competitor.

In spite of severe quality issues with the software, this next evolution of commercial GIS software would continue to lumber forward. Heavy marketing would further entrench users of commercial GIS until open source GIS software became more mainstream. At this point in time, the commercial GIS software had become synonymous with GIS as much as that Windows operating system is synonymous with "computers".

The commercial GIS software was not cheap and today's demands required best-in-class software that works with large amounts of data. This was very expensive software. It cost over $10,000 a license and it didn't even include raster functions; that's another software extension to purchase and maintain. It's quite the paradox that people feel open source software is free and therefore must not be of good quality. Anyone who has used ArcMap for any length of time can share in a number of stories about product failure and frustration.

The fact remains that ArcMap software was very poor quality software and is only marginally better after 20 years in existence. This statement by a seasoned veteran in GIS software applications doesn't need to be trusted, others can verify this assessment for themselves. The sheer number of unhappy users was astonishing, even more so was that people still had to use it.

It doesn't take long to search the internet and find forums, blogs, and web pages bringing forth an unbelievable number of complaints against the software and its vendor. One of the funnier pages is from Benjamin Spaulding's blog analysis.[22]

It was both troubling and reassuring to know that this software was affecting the entire GIS community and not just our organization. It's not a secret anymore that this commercial GIS software is awful, go ahead and Google something similar and see for yourself.

Even today, as this book is being written, there was just a post on that web site that links professionals. Someone had complained about ArcMap crashing and losing all of their development work or code. One user even commented that they "Save after every edit", or as we have come to know it, "Save early, save often". This was in the year 2022, over 20 years later from ArcMap origins. It's incredible to think just how awful this software was, even until the day ArcMap Amateur was replaced with ArcMap "Pro". It's just remarkable how expensive this software is initially and every year thereafter, while complaints about stability plagued it some 20 years later. The only parallel might be that well known commercial operating system everyone uses, but certainly doesn't enjoy.

---

[22]More on ArcGIS "Sucks" https://www.benjaminspaulding.com/2011/03/10/arcgis-sucks/

6.9 The Software Can't Be That Bad, Can It?

Buggy Software

Did you know that the term for software bug originated from an incident in 1946 where a moth was responsible for a system not performing correctly? It's true. It's taped in a log book in the Smithsonian. Anyhow, bugs in software are quite common. It is difficult to write perfect and complete code for the first attempt, but what this commercial GIS company was doing was something far beyond this.

The idea that serious flaws in commercial GIS software would persist for years without a fix, while charging $14,000 and $3,000 yearly for maintenance was completely unacceptable. One started to wonder where the $3,000 annually paid for a license was going. It was clear to users that the annual maintenance income was not going into code repair, it seemed like it was going solely towards marketing and gimmick extensions. As a user, it felt like beta software even though that testing period was supposedly over. Or was it?

There were a wide variety of bugs. A lot of research could be put into this, rehashing old commercial GIS bug stories from the past, but what's the point? We don't want to be here any longer than we have to. We want to move on to the good stuff.

The crashes were so frequent that with edit sessions we found ourselves right back to the "save early, safe often" mantra. Edit a few features, save changes, but don't get caught editing too long, or a crash could leave you without hours of work. This was shiny new software was supposedly best-in-class, or is that merely marketing spin?

> One peer from the consulting world shared how their company had to keep an accounting code solely for tracking "unbillable time due to crashes" by the software because their clients noticed it was more expensive and took longer to use the "new" software. Unfortunately, it was the only product that new Geography grads had been trained on.

So many bugs and enhancement requests had been logged over the years that eventually a request was made to have a copy of them. Surprisingly, they provided it in a spreadsheet. Oh to have that spreadsheet now; what a rich amount of information to have fun with. What would the point be though? We all get the point now, so we'll just go with one well-documented bug as an example. This particular bug was openly admitted by multiple people in tech support and even regional representatives. We are unsure of its resolution to this day.

The PDF Bug

Times changed to digital everything. Paper maps became digital maps (e.g., PDF) and those large paper field maps were going digital with mobile data collection. In the GIS field, PDF exports had to be, and must still be, the most exported map type in GIS. This PDF export bug was just one of the great ArcGIS mysteries that was unresolved for years. The exports were not anything custom or special, they were considered normal as we were using the default options.

The digital age was here and it was now common practice to print to a PDF and then review the PDF maps on the screen. The exported map product always had to be reviewed because what was seen in a software layout before printing is not always what ends up in a PDF or printed copy. Once we saw this peculiarity, we spent days trying to figure it out. Who knew how many maps with this bug had been sent out to lawyers, agencies and clients.

The mystery started with a question in the office. "Hey, why is there a copyright symbol on the map where the north arrow usually is?". The problem was initially shrugged off but when investigated further, sure enough, the north arrow had turned into a copyright symbol. Some PDF printouts in the office were noted to have a circle with a C around it. The north arrow symbol marker was being translated to a copyright symbol when exported to PDF and then printed. We didn't really have a use for printing from PDF format because we'd just print from ArcMap. Who knows how many maps were sent out before this was realized.

> Contrast this with the good old days when ARC/INFO Workstation would create PostScript files that were standard for printing from UNIX and optimized to print to plotters with limited memory. This followed the way that printing shops had been doing it for years. Instead, massive PDFs were impractical and, at the time, almost impossible to open and preview if they were too high resolution or the machine was underpowered.

Days were spent troubleshooting and calling tech support only to try different options, reboot the software, reboot the computer, load font sets and more. The volume of maps sent out was alarming. How many IT people quit their job or broke computer equipment in frustration trying to resolve why a PDF map on the screen prints with a copyright symbol in their office? A bug is maybe okay, but when a bug persists for several releases, then years after it's reported, it becomes not okay. Especially when everyone is paying their enormous annual maintenance. There were many promises by tech support and representatives that the PDF bug was going to be fixed in a subsequent release; good thing we're still not holding our breath.

There were an enormous amount of maps sent out with a glitchy copyright symbol. Wow. But hey there's a neat new gadget-kazoo to click now in a new ArcMap version, so look at the shiny new objects. That really is the trick with commercial software isn't it? Aesthetics, ease of use, or something superficial that allows users to look past muddy code. Is there not a developer who could fix this PDF glitch for at least the cost of just one year of maintenance at $3,000? Apparently not. PDF was, and is, a common way to convey maps. This unresolved issue was only one example of an unbelievable barrage of issues and problems encountered with ArcMap. Unsurprisingly, it was met with owner disregard.

Loss of Confidence Across the Industry

The buggy software impacted organizations' belief about upgrades being production worthy. It became common practice in the organization, and others, to be at least one version behind the current release. No one wanted to be the testing ground for the next batch of code edits designed to fix or enhance issues. When patches had been applied to a version and it was working for 90% of the daily tasks, it was definitely a better place to be than taking a gamble with a new release.

These bugs can't be expressed as egregious enough. Feel free to Google for more examples, but we don't want to clog this part up anymore. This was expensive software that was supposed to be top notch. If it sounds petty then you simply haven't endured the early—even not so early— years of this software that has become an infestation in our GIS community and pockets.

It wasn't just one thing that did or didn't get resolved; it was a nonstop horror ride from the early 2000s until maybe post 2015, filled with glitches, application crashes, and confusing behavior that was often problematic to pindown. These problems persisted for years without resolution with promises of repair going unfulfilled. Error messages as detailed as, "Something went wrong" seemed more applicable to the commercial GIS software in general. Something went wrong alright and it was this darned commercial GIS company in the pursuit of profit over quality. GIS was a cool technology and they were ruining it.

Plotting

Before mobile data collection was in regular workflows, large 3ft x 3ft maps were plotted and used in the field to mark up for land use. The project area of 20,000 acres required some 40 maps at this size and scale. These maps were easily automated through scripting in ARC/INFO ArcPlot with clean leaders and auto labeling. In ArcMap, these auto-labeling features were gone. Nice. Same price, but a loss of features? The labeling in ArcMap was so bad that a third party company built and sold an extension to address labeling problems and called it Maplex. The commercial GIS company eventually bought it and incorporated it into their software - who knows what tier level has access to good labeling.

Not only were the cool auto-plotting functions gone, but ArcMap often failed to plot altogether or sometimes just partially. Many calls to the company were made and there was no resolution. It was escalated to who recalls what tech level or representative level. After a relentless pursuit of what the problem was, eventually it was stated that with the ArcMap re-write, wait for it, there was heavy reliance on the Windows operating sub-system for printing or plotting. In other words, it's not their fault, it's the operating system's fault. There's no way to "fix" this one.

This sounds strikingly similar to the problem of erradic networking and data loss between an Arc/Info Workstation and Windows NT server. This, time, the blame was on some other company, and not just some other company, one of the world's largest software giants. What to do? After this astonishing admission by the commercial GIS company, and without any customer help or support, we called the Windows operating system company. Talked to specialists who had no idea what we were talking about. Nothing; there was no known plotting error on Windows. It was a dead end.

As bad as this problem was, it's too bad the details cant be recalled. The proposed solution was even more absurd than the problem itself, which was to—if you pause you might be able to guess it—buy an extension. The extension was called ArcPress and it was an extension that rasterized a map before plotting it. PostScript has very high detail print capabilities and uses coordinate descriptions to draw vector features. It was the preferred printing and plotting type that now couldn't be used. This problem was a major hit to production in time and materials lost.

Commercial Software Performance

There's not much to say here other than it's not good. ArcMap is a 32-bit Windows application. Modern computers run on 64-bit operating systems and have for over 20 years in regular use. When applications like ArcMap run in 32-bit space on a 64-bit operating system, the applications are bound and limited to 32-bit performance. There are still 32-bit limits on RAM and resources. Meaning a 64-bit operating system does not help ArcMap run any faster, other than handling secondary system resources that might impact ArcMap.

It is a major sacrifice in performance to be operating in 32-bit space. GIS is resource intensive software and the performance only gets worse over time as demands on the software as the data becomes more dense. Data resolution only increases with technology and time. LiDAR is an example of data that has evolved with technology and is easily in the billions of data points per work area. Aerial photographs are now sub-foot pixel resolution and the resources to handle these dense datasets is not sufficient with 20-year old software. Of course, spoiler ending for the commercial software section, they have to do another rewrite of a *professional* or *pro* version called ArcMap Pro. I'm not sure what it says when you have to call your software professional, but doesn't this imply the Arcmap non-pro, the 32-bit version, was well, *amateur* at best?

It's as if some software makes one a professional. Faster computing power is needed to deal with the large datasets we work with today. Open source GIS is 64-bit software, often compiled for the specific machine it will run on. Was it mentioned that the performance of open source software is among the best there is? Yeah, we discussed this; it undoubtedly gives a performance advantage to open source software, it's not even close. Another win for open source.

Tech Support

In the late 1990s, the commercial GIS technical support was really quite good. Early on, tech support was called for "how-to" knowledge and the calls were answered by intelligent people who understood the software and how it worked. As the commercial software users grew to new levels, it became clear that the goal was to provide a quantity of technical support, not quality. Over time, with ArcMap saturation, the quality of tech support plummeted.

Talking to tech support over the years provided a lot of useful insights. Every year, there seemed to be an ArcMap release or update. The releases didn't seem to coincide with a new function or development of code and users could only get the updates if they paid their annual maintenance. One somewhat

surprising revelation by tech support was that software updates were provided to users each year, just to make users feel warm and fuzzy about getting
something in return for the enormous maintenance fees they paid. While this
was always suspected, there it was in full admission.

Some additional insights were gained through casual conversation with commercial tech support and representatives. Open source GIS software was often
talked about as a comparison to issues occurring with the commercial GIS software. When there was agreement about why open source GIS was better at a
particular thing, there was more than one occasion where the reply would be,
"Okay, but there's no money to be made off of open source software." These
tidbits of information over time only began to reinforce and shape the idea
that commercial GIS software was a hoax, not interested in providing quality
software, only interested in profits.

The open source community had a successful technical support system that
was based in the use of mailing lists. Users relied on each other, the community, to answer questions and solve problems. What better way to offload
users from your commercial technical support staff than to capitalize on this
free, community-based behavior. Rather than mailing lists, the commercial
GIS company strongly encouraged users to go to forums to get help or to log
product enhancement requests. There was more than one time technical support staff tried to direct us to these forums. Why would someone use a community support forum, when paying thousands of dollars every year for the
ability to pick up the phone and call tech support?

Using open source mailing lists, it is entirely possible to encounter the developer who worked on the module in question. With commercial tech support,
this is impossible. It was stated that not even tech support has access to the
developers to ask questions. Putting a wall between technical support staff
who encounter user issues and developers seems like a good way for the two
to never cross paths. Open source wins in this category of tech support.

Corporate Representatives: No Help

There was no resolution with tech support for some of the major issues. Quality support and software were long gone. Constant replies of "it's in the queue
to be fixed" were heard until it wasn't even believed. ArcMap Pro was coming
out in a few years, a new software to fix all these woes. Who would believe it
after all these shenanigans.

As a result of mediocre technical support, some issues had to be taken up with
regional and east coast representatives at the highest levels. These concerns,
which were valid and hindering production, were met with disregard. Con-

tempt was just oozing from these representatives, with comments that mocked and taunted the user, "Well, what other software are you going to use?"

Due to the frustrations with the commercial GIS software, open source GIS was being explored as an alternative. When open source GIS software was mentioned to corporate representatives, they just laughed at the comment as if open source GIS software was a joke. Corporate representatives quite obviously don't exist in open source communities. They're not applicable and are just another cost that needs to be dug out of users' pockets.

6.10 Software Switcheroo Complete: Minus One User

The Big Switcheroo was complete. ARC/INFO Workstation and ArcView 3.x were dead. ArcMap had emerged on the scene while users absorbed it in stride. Complaints about this software quality and the exorbitant annual fees were growing in the commercial GIS user base. A large number of people were dissatisfied with commercial software. With not many apparent alternatives, all users could do was complain. There's a whole section of the Internet allocated for this activity. Eventually, some of these users would become open source GIS users and leave the commercial GIS toxicity behind.

The problems that plagued ArcMap were so systemic and pervasive, compounded by the care-free and insulting responses by the commercial software company's representatives, that a new user had been born. One who had a point to prove by taking this frustrating and unhappy existence with commercial software and making something awesome out of the situation. There was a crucial decision to be made at this point: leave GIS software altogether and find another career, or find another GIS software. After all, it was just software and learning to use software was a speciality that was nowhere near over.

There would be no more commercial giant software fights. No more calls about issues. No more caring. Just accepting the fact that commercial software only loved the dollars behind the GIS technology, not the technology itself and certainly not the users. There would always be a group of fans for this commercial GIS software, mostly out of not knowing any better or even to go exploring. It's with great hope that some open source GIS users will emerge after this book.

This was it, the abandonment of commercial GIS software. Sure, the commercial software would still be used out of an organizational need, but there wouldn't be any fawning over the monopoly giant. There could be no love given after the realization that all they chased were profits. There would be a rebellion, a resistance to use commercial software every step of the way while always turning to open source software. Stubborn maybe, but the preferred word is persistent; and wow did it pay off in the long run.

6.11   Keep It Fun

The fact is, GIS software is fun, really fun. The tools contained in the software allow for many creative applications. Solving real world problems sounds so cliche, but it's exactly what the software has the capability to do, provided the user has the creativity to think around corners. It's disturbing to think that there was serious consideration to exit this open source GIS journey because of commercial GIS problems.

When things get rough, dig deep. Keep the faith. Never give up on what you believe in. These cliches that have evolved over time are really something worth noting, because it turns out this desire to stay the GIS course, but not down the commercial software path, was really worth it after all. There are no regrets for the path chosen on this journey.

7. How Open Source GIS Saved Me

7.1 The Joy of GIS Returns

Enjoy the Journey, Ignore the Destination

No good story is without some initial suffering to get to a better place and wow, there was plenty of suffering! The old "lemonade from lemons" cliche comes to mind. So, *good bye commercial software*; wish it could be said it was fun. It's here we give commercial GIS software the boot.

Up ahead, we've got web mapping, hydrologic analysis, LiDAR, Enterprise GIS Databases, Minecraft and even a theoretical geoblockchain designed after Bitcoin; all with open source software. Not unlike the time rushing back to meet the carpool that ended with a ticket and broken clutch, we'll slow down and soak this good stuff in. Ralph Waldo Emerson is credited with the quote "Life is a journey, not a destination" and there is certainly agreement in the idea that the joy is in the journey not the destination. This part of the journey was worth slowing down and absorbing.

As a result of the constant frustration with commercial software, there was a challenge and drive to find another GIS software solution. Linux was already being explored as an option to Windows, and interestingly enough, for the same reasons: frustrations with commercial software but at the operating system level. Linux exploration at home led to web serving and other activities. Linux was nerd stuff and this nerd was ready to fully embrace it.

Since this whole journey is based on memory, some of the official dates and timing cannot be for certain some 20 years later. How could one possibly know that a book would be written some 20 years later? We need more stories to share with people, so maybe you'll write one someday. Be sure to take good notes along the way.

The Open Source Community

After high frustration levels with commercial GIS software, there came to be extreme disregard for that commercial GIS company and the eject button was pressed. A soft landing was found in the receptive and open (no pun intended—okay maybe) nature of the GIS community in the open source world. The mailing lists were a great way to communicate with many people in the community all at once. The users were extremely helpful, knowledgeable and incredibly smart.

Alone, users can know some stuff, but in a community like this, the living hive-mind can adapt and solve any problem. The community has everything from casual users to experienced users, new users, developers, idealists and very few fools or people who can fake it like with commercial software. The community acts as a collective and reacts similarly, yet each having their own identity. This was the reason open source GIS was indestructible and would never die. Less intense, but not unlike what has become known as "Bitcoin Twitter"; the Bitcoin crowd on that popular social short-messaging application.

Open Source Geospatial: Let's Go

It was here, jumping into the open source GIS community, that no curiosity went un-investigated. A great deal of time was spent learning and checking out all the gems in the Geospatial Free and Open Source Software (GFOSS) communities. A new supervisor had been assigned, maintaining the winning streak of cool bosses. Open source GIS success could not have been as much without his support and engagement with water resources.

The order of the origin story might be slightly off, but it all started happening at once. The more time that was put into open source software, the greater the returns were. Sure, the same can be said about other software, or mostly anything in life, but open source is different because users have to pursue it. It's not like a commercial certification course or software extension that levels up a user, open source users have a driving force from within to learn more. Engagement with the mailing lists and software begin to show their rewards; learning, growing and expanding skills as fast as one can absorb them.

Without sounding too hung up on philosophy, which we really mean as a general sense of belief, nothing overly sensational, the open source nature of the projects is what makes them special. Learning about the open source philosophy and adopting it was perhaps the most important knowledge gained in this adventure. A close second would be seeing other viewpoints from around the globe. Viewpoints not entrenched in the commercial GIS marketing propaganda that US users were faced with. Maybe these are things best appreciated as time goes on with a journey.

The circle was now complete. The story originated in university with GIS on UNIX, all the way through the miserable existence of GIS on Windows, back to GIS on Linux. For over 20 years, GIS on Linux was used for analysis, modeling, web applications and to create an open GIS database at an enterprise level. It was time to send this trip to the moon, leaving commercial software for the earthbound suckers. So sorry, *suckers* isn't quite the right word, maybe *unknowing* is better. Can't know all this stuff about open source software and feel any other way that people were tricked; including the rider at the origin of this journey.

In the GIS Software User Minority Now

One of the first things realized when joining the open source GIS software community is just how much commercial GIS has infiltrated the US market place. It's everywhere. The early part of the journey consisted of being in the GIS software majority. It was apparent the commercial GIS company could write mediocre GIS software at best, but they were true masters of marketing their software into every nook and cranny of the US market.

The environmental consulting firm was committed to commercial GIS software for many reasons, some valid, some not. There is no doubt these factors played a role: marketing propaganda advertising "the best", client compatibility, and bidding on projects that require the commercial software file format. Ah yes, this closed circuit is quite apparent now, isn't it? One must have the software for bids and client "compatibility". Not really, but no one would believe otherwise.

Why would clients and bids request a special commercial GIS file format or software? Okay, maybe file format for compatibility, but to say people must have commercial GIS software really highlights the ignorance we've reached in GIS. It's just points, lines, and polygons with some attribute data. It doesn't have to be tied to a commercial software brand, but alas, it was so. Clients and bids requested commercial GIS software and if you didn't have the software, don't even bother applying to get the GIS work.

Once one leaves the GIS commercial software fanbase and turns to open source software, it becomes apparent just how much work excludes open source GIS software. It became concerning how many *Requests For Proposals (RFPs)* had *requirements* for expensive commercial software. We can certainly get data in a GIS format and not care about what software created it (see "it's just software"). Whenever the use of these data formats was questioned, there was never a good answer or defense as to why proprietary formats were in the mix. The propaganda was far too strong in the United States to allow even open-minded people to consider alternatives, or even think beyond what they've been told.

Linux in the Workplace

Use open source GIS software on a commercial operating system? There was no fribbon way Windows was going to be used for open source GIS so we had to get Linux into the workplace. Linux in the workplace had to start somewhere and it began on upcycled, old PCs slated for the trash. It would be great to say it started with great corporate support and was fully embraced, but the fact was, it was a threat to the Windows guy. Remember the lesson earlier about the IT friend or foe? This one was a hard-headed lesson that hadn't been learned yet, better to include and befriend them for sure.

There won't be an argument or debate; we're just going to go with the opinion that Linux is better than Windows. Period. There was one last contentious situation with a person in charge of IT and it was all because of Linux. People who are really into computers often become aware that Linux is superior to Windows operating systems. If "computer people" don't know these facts, then they can be threatened by what they do not know in just the word Linux. After all, they also have been inundated with messaging that said Windows was the best and to question it often put them at risk of being on the outside in the IT world.

At one point, there was a myth spread that "The GIS guy is trying to take over the company with Linux" or something of that sort. It was a constant uphill battle until the new, open-minded, IT person came aboard, the one who termed the work as "environmental modeling".

Back to School

This home hobby with Linux expanded into work and became an obsession. The software, the community, and the potential with geospatial software were all too enticing to someone who had just made a crucial decision to abandon commercial software and find something else. After learning as much as possible about Linux, it was time to fill in the gaps with formal education.

There were times that code needed compiled and this was definitely not the click-and-pick stuff one finds in commercial software. There were GIS web-based mapping services that needed special configuration and this required understanding details about the operating system and web server. These were nerdy, specialized computer software packages that were not taught with GIS. The local college had some excellent courses in Linux administration and the professor was top notch. So it was back to school for *Linux Administration*.

These classes were no disappointment. There was so much to learn about Linux from a knowledgeable professional in an academic setting. The courses were a lot of fun and really interesting, so much fun that more courses were taken. Perl programming and network security classes proved to be very useful with Linux. It was during these classes that another cool professor was encountered and later developed into a lasting friendship. The Linux administration, Perl programming and network security courses created a solid background to administer a Linux workstation and provide web mapping services. Now knowing a little more about Linux, it was time to get immersed into GRASS GIS.

GRASS GIS: A New Hope

Back to GIS on Linux. This was awesome. Knowledge about Linux was growing and so was the knowledge about GRASS GIS. There was a new curiosity that could not be satisfied as the knowledge about Linux and GRASS GIS was growing. There was great interest in the community of users too. Users ranged from native open source software users to commercial software converts, although there really weren't many of the latter.

In the beginning, GRASS GIS had an initial strength with raster data and analysis, while the commercial GIS company had its strengths rooted in vector data. Yes, there are many opportunities for puns with roots and GRASS GIS, but we won't keep going there. GRASS GIS isn't just a cool name, it stands for *Geographic Resource Analysis and Support System*. We won't get into historical details, other than to say it was built by the *US Army Corps of Engineers* and later released to the public. From there, it was developed further and licensed as open source software. Over the years, the vector management and editing have evolved to top notch quality.

When joining the mailing lists, it was interesting to observe that there were only a few users from the US. Most users were from Europe or more generally, not from the US. Why was this? It is no mystery at all. In fact, it could be easily argued that there was an active campaign to squash and suppress any other GIS than the commercial brand that dominated the US for so many years. Active and passive campaigns, complete with lobbyists, made best efforts to ensure that their commercial GIS software was placed in front of unknowing students, corporations and government agencies.

Topology Implementation and Management

Topology helps maintain GIS layer integrity, but commercial GIS only gives users these tools if they have deep pockets, to the tune of $7,000. This is criminal. Why would data integrity not be something considered important or valuable at all tiers of commercial software use? Overlapping and gapping polygons are a serious issue for GIS data; they make data summaries wrong and inexact. We can say with a high level of confidence that data integrity was not as important as taking more cash from users. What a joke, you can only do topology checks for $7,000. The entire community of GIS users have to use this data, so why not provide tools for good clean data in all software tiers? That would be a move in the interest of the community and not profits.

On the other hand, GRASS GIS has topology integrated with editing because open source software isn't configured for pay-to-play schemes. The data structure uses special closed lines called boundaries, which are really polygon edges. Contained within these boundaries, we have a feature called a centroid that stores the attribute information of the polygon. Boundaries with nodes and centroids can be seen and manipulated in an edit session, however, outside of editing they appear as a typical polygon with a fill and outline to symbolize. When building topology, the software reports any topological errors that the user may need to address. These errors are color coded in an edit session so that users can quickly identify and fix them.

Easy Transition to Open Source GIS

There was an immediate affinity towards GRASS GIS, because it had an interface similar to ARC/INFO Workstation: a command line terminal with a (sorry) somewhat crude GUI. How ironic to not only return to GIS on Linux, but to return to a *very* similar sophisticated GIS software on Linux. Now that's *ha-ha* funny. Later, the GRASS GIS GUI would be appreciated for simplicity. The GRASS GIS GUI was better than ARC/INFO and the command line was totally scriptable in a wide choice of programming languages.

Initially, when Linux and GRASS GIS were encountered on this journey, there were some difficulties getting the software configured. It was not as easy as it is today. For users who knew Linux pretty well, installing geospatial software would be no problem. For users who didn't know a lot about Linux, they would have to learn Linux and open source GIS software simultaneously. This sometimes proved frustrating because half the time could be spent trying to get an extension to work rather than doing GIS analysis. There was more than one time the realization came, "Have to stop messing with Linux and get some GIS work done!"

There were huge benefits that came with learning Linux because it was so tightly integrated with geospatial software. Tools like awk, sed and bash scripting all proved to be beneficial with GIS going forward. There was a lot of information to learn, code to be compiled and libraries to be linked. There were struggles with Linux that would be taken for granted on Windows, but when everything was working, it was consistent and solid performance. Untouchable. This would be a recurring theme, things that typically worked, more or less, out of the box with commercial software might require some additional software or knowledge with open source.

Of course, what we're really talking about here is a lack of knowledge, not a shortcoming of Linux or open source GIS. Not knowing about an operating system or software doesn't mean the software isn't any good, it just means the user has a lot of catching up to do. And a lot of catching up was done. Quickly. Once down the rabbit hole of open source GIS, a lot of time was spent learning about all the details of GIS that are hidden with commercial software. Learning what buttons to press is important, but it's the knowledge behind the buttons that matters.

Hydrologic Modeling Revisited

Here we were. The time had come, once again, to chase the dream from university: combine hydrologic studies with GIS software. The first attempt with commercial GIS software did not go very well, but with a change to more sophisticated software, the outcome would be much different. This open source software was really hitting all the key points: very technical, running on Linux, open source, reliable and high-performance. What a high contrast to the last experience with commercial software.

As knowledge about open source GIS software grew, so did knowledge about streams and the local terrain. Open source software required knowledge about how things worked more so than the button-pressing solutions offered by commercial GIS software. Doing the best job possible meant learning about the science behind the technology, which was local hydrology. University was in New Jersey, but we were working in central Florida now. The terrain differences were very different and the differences are worth noting as it relates to hydrologic modeling.

Background on Streams From a Non-Hydrologist

There's no hydrologic expertise here but we're going to pretend. Streams are where the water table intersects the land surface, the same with lakes and wetlands. There's other conditions that are exceptions, such as perched wetlands or ephemeral streams, but the groundwater-land intersection is the rule. Streams that we see and study in most places are recognizable and well defined. They create a drainage pattern that is very recognizable and are formally known as dendritic drainage networks (Figure 7.1). The streams have a valley in the low areas and a top of bank along the highest sides. Stream geometry is carved into the terrain over millions of years and is moved around by erosion and deposition of sediment. Streams can have perennial (year round) or intermittent flow.

[1]

Figure 7.1: Dendritic Drainage Network

Highly disrupted stream networks are where the streams don't have a continuous connection (Figure 7.2, on the next page). Deranged drainage networks are the most extreme form of this, where historical glaciers completely disrupt the current overland flow. What can cause a stream to become interrupted or vanish? The best example showing how a stream vanishes is when it enters a lake. The same idea applies to flat terrain and topographic depressions such as wetlands. Some of the stream characteristics—the banks—can lessen in flat terrain as the water velocity slows down.

In topographic depressions, often wetlands, the stream geometry will vanish entirely as it enters the depression. The stream often reforms when exiting the depression, as the water starts to flow more rapidly and carve the landscape again. The resulting dashed line network from these interruptions may be easy to see in the example (Figure 7.2, on the facing page), but rest assured that in reality, these disconnected streams are not as easy to find or identify.

Hydrologic Modeling Challenges in Florida Terrain

Why do we care about Florida terrain? One reason is because it's unusual, the second is that it's where the story takes place and three you might have to cope with a similar terrain one day. For most of Florida's history, it was under

Figure 7.2: Disconnected Streams

water, then about 2-4 million years ago, it would slowly become a land mass.[23] Florida terrain was primarily shaped by marine influences and, in terms of landscape age, terrestrial Florida is relatively young. Vast areas of flat terrain speckled with a high density of topographic depressions, wetlands and lakes, are very common. There is an area in central Florida, called Bone Valley, where the dead ocean creatures create a great phosphate mining environment. Lots of shark teeth can be found in the deep mine pits.

Among the three terrain types, Karst, Highlands, and Flatwoods, the last is the most challenging to model with regards to overland flow. Some flatwood areas are broad and shallow, making stream channels hard to delineate. There are sink-watersheds, drainage areas that drain through groundwater. Wetlands are everywhere and also disrupt stream continuity. Due to Florida's relatively younger surface exposure, surface flow has had less time to shape how the water flows across the land in comparison to other states.

7.2 Hydrologic Modeling in GRASS GIS

Whoa. <insert Keanu GIF>. As GRASS GIS was explored further in the early 2000s, it was discovered that it had a whole section devoted to hydrologic modeling. Now this was good stuff. Excitement was bubbling up and GIS was fun again. There were a lot of cool hydrologic modules that could do advanced modeling. First let's check out the raster-based GRASS GIS command r.watershed.

---

[23]Prehistoric Florida, 12,000 years ago: https://discover.hubpages.com/education/Prehistoric-Florida-12-000-years-ago

r.watershed

r.watershed is a powerful and well designed hydrologic modeling tool with some neat outputs, but the most used ones were: drainage area delineations and stream network extractions. The most important one was the drainage area delineations because they had regulations about pre and post development, although the later example will show the value of the stream accumulation outputs. Can't recall the tolerance for drainage area sizes, but they did need to have the same Strahler order, pre and post development.

The commercial GIS software had issues with depressions or even a single pit in the terrain. The user had to fill all the depressions until a direction was determined for every cell. None of that was necessary with r.watershed. It could cope with sinks and pits just fine. The algorithm was intelligent. If flow accumulation found its end in a sink, then r.watershed would route accumulation out of the depression and keep tracing the lowest terrain.

There were a lot of hydrologic modeling for various projections going on. While this work was going on, code in r.watershed was being updated by a developer. The module enhancements were posted on the mailing list and questions were asked to the developer. The next part was cool. As patches were applied to the code from the developer, code patches were sent to test the pending features. It's a great feeling to offer any help to developers, as they are the reason we have this great software to work with. There's always a way to help the open source community; just listen, wait, and then hop on in.

Anticlimactic Success: The Path to Unstoppable

The time had come to revisit the initial drainage delineation project, about 20,000 acres, with this new software and operating system. It was time to fire up r.watershed and do this. There was no LiDAR data to use at the time, so two foot, pre-development contours were used to create the DEM. r.watershed was then run to determine drainage areas as well as an associated drainage network.

This time, everything worked as expected. There weren't any software crashes. There weren't any problems due to a non-hydrologically conditioned DEM. There weren't any apparent errors in the output. It looked great. The basins were delineated and flow accumulation lines, illustrated as streams, were also delineated.

It worked, but was it correct? How could we know? There were drainage areas delineated by an engineering firm and they were considered the best available. Let's compare the GIS outputs against the other basins, and wait for it, the modeling matched. Down the road, LiDAR would make the GIS modeling superior, but for now we knew that open source GIS modeling could compete with engineering quality delineations.

r.watershed did a great job delineating the drainage areas, but it also had another output that was useful: accumulation areas that represented streams. Previous drainage area models by engineers did not display a flow accumulation along the lowest elevations. Their model outputs had connecting links from one basin to the next. The GIS product was far better overall, because the natural drainage network exactly matched topography, and the corresponding drainage areas exactly matched the drainage network.

This success was the beginning of something greater to come. A bit anticlimactic, but all it really takes is a success story to start building on. This initial success was the boost needed to advance to the next level. The embers that died down after the commercial GIS software flop, were rekindled with open source software. It was full speed ahead. The dream from university to the workplace had come true. A passion for open source GIS software on Linux grew ever more insatiable as the quest for more continued.

7.3 Web Mapping Goes Mainstream

What is Web Mapping?

During the same time as the hydrologic modeling take off with GRASS GIS, web mapping was encountered. Web mapping is the intersection of GIS and web technology. The data is distributed through a web page. MapQuest was one of the early examples and now we have Google Maps. You get bonus points if you remember MapQuest's aided fall from glory. Web maps have a scale bar, north arrow and legend, just like our old paper maps. Web maps can be a great user experience because they provide interactive, streamlined and simple GIS functionality. For example, pan, zoom, query and measure are common, but not editing, overlaying, or calculating slope. The customizable nature of web maps makes them highly configurable for specific end user needs.

There is more to web mapping than the appearance of a map in a web page. Behind the scenes, there is a web server, geospatial data, and geospatial web software. The last part might even be gone today, with how flexible programming is. It might be fun to know that while MapQuest came out in 1995, the Web Map Service (WMS) open specification didn't come out until 1999. This allowed people to build web maps with a common specification.

Way back in the beginning, we covered the geometry of GIS features. Not only can there be hundreds of thousands or millions of features in a layer, but the line and polygon features can have many points that define them. Using this data on a local workstation isn't a problem and using a shared data server in a LAN configuration with workstations isn't a problem. However, once we want to share data layers outside an office—across a WAN or the Internet—we need to format the GIS data in a web-friendly format. A format that doesn't have to deal with millions of points and vertices defining features, resulting in a crashed web browser.

A clever way to solve this problem is to make an image of the map on the server side and then send this very tiny, yet high quality, image to the end user's web browser. The browser interface understands the coordinates as geocoordinates, so users can interact with GIS layers similar to a desktop GIS. Navigating large datasets is very manageable and seamless to the end user. Web Feature Service (WFS) technology sends vector features to users, but filters data by extent and does it very efficiently. WFS features can be interacted with like a normal GIS, but WMS "features" are just an image.

Use Commercial Web-Mapping Software? Ha!

It was probably somewhere around 2004 now. The benefits of open source software had been realized through hydrologic modeling. At the same time, there was an interest in web mapping technology. ArcIMS was the commercial GIS company's web mapping software and a trial for the commercial web mapping software was set up by the local IT representative. This commercial stuff was really expensive and really clunky. Something like $10,000 for a one processor system and $20,000 for a two processor system. Wow, another $10K per CPU? What a gimmick. Don't forget the annual maintenance costs. What could we do with open source geospatial software instead?

One of the things the late 1990s tech bubble brought us was mainstream adoption of the World Wide Web. It was inevitable for web technology to come to geospatial. At the beginning of this intersection, there weren't a lot of choices in web software like there is today. One of the first applications for web mapping was using a *CGI-BIN* configuration. This setup worked incredibly well on Linux with Apache web serving software. The first web mapping software used in this journey was *UMN MapServer*. This was a very tiny, yet fascinating piece of software from the University of Minnesota (UMN).

Web Mapping Illustrated, by Tyler Mitchell

Hey, we know this name, great guy! During this adventure, knowledge about web mapping with open source software was craved. Mailing lists were a great way to exchange current information and topics. Current information is actively shared on the mailing lists, where the information flows and is absorbed much better while it's happening live. If a user was searching for historical information, the information was available, but users would have to dig through old email threads. Desiring a cohesive presentation of web mapping, the first of a few open source books was purchased, Web Mapping Illustrated by Tyler Mitchell.[24]

One of the cool things about the mailing lists was the people encountered along the way. Tyler was one of the main players in the open source revolution and it was really cool to interact with him on mailing lists. Then to top it off by reading his book. Awesome. What a tight knit community of geonerds. Connecting with people in this way helps build a strong user community, a sense of almost "knowing" people you don't really know. Some 20 years later, Tyler would make a suggestion to write this book. How cool this overlapping story is when taking a glance from the big picture, or small scale for you geonerds.

Web Mapping Success

Time to put web mapping into practice. We had an ongoing project that evolved into a large group of consultants, from different companies, working together for a client. While areas were being reviewed with the project team, it was becoming difficult to address the dozen layers. Desktop sharing wasn't nearly as popular or easy as it is today, so we needed a tool that could share GIS information among the team members.

The local IT guy was trying to gather support around his ArcIMS demo, but that was deader than dead now that open source software was here. UMN Mapserver was chosen to be the GIS web software and it used a CGI-BIN configuration to serve the GIS layers. There was a steep learning curve, but the software was much nicer than ArcIMS. We used an old PC to install the server software and then distribute the project data over the Internet to team members. It was a home run again.

---

[24]https://www.oreilly.com/library/view/web-mapping-illustrated/0596008651/

Money Appreciated, Tech Not So Much

Much of the problem going forward was that there was little to no organizational support for these highly technological efforts using Linux. The organization had a large number of experts in the environmental consulting industry, but not in the way of technology. This lack of understanding or interest in technology would become a serious issue going forward. New technologies were brought to management, but met with blank stares. One technology inquiry to a supervisor brought the response, "I study dirt." with no further follow-up. Ouch. Tough crowd.

Lack of interest and engagement turned out to be harbingers in retrospect. Web mapping technology wouldn't be used until some 15 years later, but at least we had one good run in the early 2000s. There wasn't an interest in the web mapping technology because there was no creativity for how to bill it. The GIS technical person wasn't a good salesperson which compounded the issue, but it seemed obvious that the cost savings and technology offerings should speak for themselves. There certainly was an appreciation for the approximate $30,000 the project brought in using open source web-mapping, but there was no encouragement to build on this success. This technology would die here in the early 2000s, way before ArcOnline and web mapping were commonplace.

Ease of Use With QGIS

In the early days of web mapping with CGI-BIN, users had to move GIS files, reset link paths, and do a lot of manual setup on the web server. Open source web mapping would evolve over the years, such that symbolized layers in a QGIS session can be exported as a collection of web files. Prepare data in QGIS, export to web server files, put files on web (cloud) server, done. There is a plugin that allows for this efficient workflow called *QGIS2WEB*. Some of the output options are Leaflet, OpenLayers or Mapbox GL JS.

This once obscure web technology was now commonplace in open source QGIS and it was absolutely fantastic. Several web mapping applications have been deployed to a Linux cloud server with easy integration and setup. Setting up a Linux cloud server, web serving software and a web mapping application can now occur in less than an hour.

The Cloud

What is this obscure word or place? About this cloud matter, the cloud is just using someone else's computer. All this hype and propaganda, "We're going to the cloud!" like it's some kind of magical place where all computing problems go away. In the early days, there were machines serving web pages, but as time went on, web servers and services turned into slices of machines inside machines. That's it. The big secret is out on the cloud now. It's just using someone else's publicly facing computer, usually slices of larger computer resources known as *virtual machines*.

This Question Again?

The first role as a Linux-based GIS Web Administrator had come and gone in the early 2000s. As an aside, is a GIS Web Administrator a GIS Professional? Are we really back to this pointless question? Web mapping applications are rarely ever a GIS, so how does it really fit the *GIS* Professional label? Web mapping is a geospatial technology, but it's definitely not a GIS. Sure you could build one, but then you should probably just download QGIS instead. These are the gray areas we get into about technical applications that have a geospatial component vs. geospatial applications that have a technical component. You decide if you feel it's really worth it to define people with labels.

Afterburners Ignite

These two notable events, hydrologic analysis and web mapping, put the afterburners on the open source journey. We had solved the initial problem encountered using commercial GIS software for hydrologic modeling, by replacing the software with open source software. The web mapping adventure created interest in geospatial services beyond desktop analysis and modeling. The journey was like a sponge and there was an infinite amount of open source software to absorb. It was nerdy and it was awesome.

As time went on and open source software was applied further, a real proficiency was gained using both commercial and open source software. As the journey was traveled, both softwares were in constant use and comparison, and it was clear that open source GIS was the better choice. Being able to use commercial and open source systems was a great advantage. Simple vector and client needs were accomplished with ArcMap and geodatabases, but any difficult or big data problem immediately went to open source software. Open source geospatial software was doing all the hard work.

When comparing, there were some tasks that commercial GIS software couldn't even do, namely multiple flow direction overland flow routing. In another client request, we would have had to purchase a commercial "Network Analyst" license, but it turned out that all of the network analysis was able to be done using GRASS GIS. Why buy another commercial extension, when open source offers it for free?

LiDAR Technology

When *Light Detection and Ranging (LiDAR)* data was discovered, it took terrain modeling to a new level. This data was free and could be used to create highly detailed terrains in GIS software. GIS is amazing software, but the quality of the data determines how awesome the output will be. LiDAR data is very detailed data and of very high quality. There was an incredible opportunity that arose in the workplace, as a team of engineers were contracted with a regional agency to perform floodplain modeling. This curious "new" technology, Li-DAR, was encountered, explored and then used to fuel the next level of open source geospatial software success.

This is where not only afterburners ignited, but alas, what's a good analogy here, not sure, but the fact is, LiDAR was responsible for a few upcoming and significant events. LiDAR data is survey grade point data that provides accurate elevations in X,Y,Z format with a high fidelity. The sensors can also capture other characteristics such as intensity and color. DEMs created from LiDAR have much value because of the detail LiDAR provides. LiDAR scanners also capture trees and buildings, which can be integrated into surface models.

We should probably explain LiDAR a little bit before we get into how it has been used. After diving into this marvelous technology and exploring the capabilities, it was time to cool down and share the information. An article was posted on that well-known professional networking site and is modified below.

7.4 Intro to LiDAR

LiDAR - Light Detection and Ranging - is such a beautiful thing. In this context, LiDAR points are obtained from an airborne scanner and have a very good spatial distribution of points. LiDAR points can be used in many applications. LiDAR captures the ground elevation, as well as vegetation, buildings and anything else above the ground level. This full scan is known as a point cloud. Software codes the points with different classifications, which can be extracted and used in GIS.

One common model built from LiDAR data is the *Digital Elevation Model (DEM)*. These models use only ground points, creating a model known as a bare earth model. *Digital Surface Models (DSM)* contain data points for everything from the ground up, such as buildings, trees, electrical transmission lines and tow-

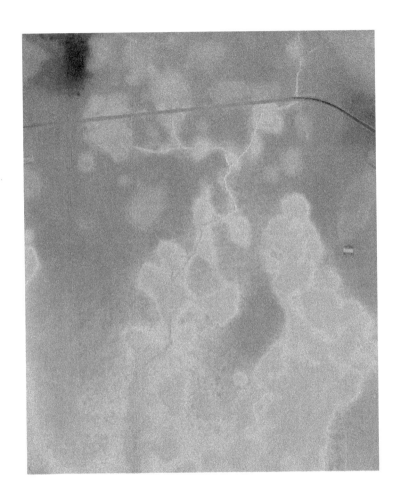

Figure 7.3: Shaded Bare Earth DEM

ers. *Digital Terrain Models (DTM)* are not as common in public distribution, as they forcibly incorporate features like roads and streams into the DEM.

Topographic Contours: Better Than Nothing

In a time prior to widespread LiDAR data use, topographic contour lines were the primary data sources used to create a DEM. Topographic contour data contains highly packed elevations along the line, with no data— nulls—between the contour intervals. This data is far less than ideal to create a continuous DEM, compared to the nice distribution of data points that LiDAR provides. Contour line woes are made worse in flat terrain because the horizontal distance between contour lines—data points—is large. Big gaps in topographic elevation data means that micro-changes in the terrain are lost. One feature lost is shallow ponding areas between contour intervals.

It can be tricky to develop a nice DEM from contour lines. In addition to concentrated elevation points, sometimes there are artifacts, or odd shapes, which can arise from too many void areas in the input data. Five foot contours are available for the entire US and in some places there can be two foot contours. These are better than nothing, but LiDAR data is always the preference to build a DEM.

Efficient Software and Powerful Hardware

One problem encountered while working with LiDAR data is the high volume of X,Y,Z data points. Remember that 20 year old commercial software stuck in 32-bit space? Turns out that it's not a good choice for processing billions of points, or billions of anything for that matter. 64-bit Linux systems running 64-bit geospatial software are an excellent choice for processing big data; they are fast, reliable and durable.

A very popular LiDAR software at the command line is LAStools; the author, Martin Isenberg, honorably mentioned. He built some GUI tools for QGIS. LAStools provide a very efficient and fast way to filter LiDAR files and can easily be scripted to process large volumes of LiDAR points. It's not uncommon to deal with hundreds of millions of points, even billions, generating ASCII X,Y,Z files several gigabytes in size. Data of this magnitude requires highly efficient software and powerful hardware to process the points in a timely and reliable manner.

LiDAR Point Clouds

The LiDAR file that holds all of the points captured by a laser scanner is called a point cloud. LiDAR point cloud data is typically stored in a binary format with the extension .las, which is short for laser. This is an open format by the ASPRS (American Society for Photogrammetry and Remote Sensing). Note the important word "open" in the previous sentence. In the US, these .las files are usually 5000x5000 feet or about a square mile in coverage. When large extents are covered, many LiDAR tiles are required. The LiDAR files are referred to as tiles because they are almost always part of a larger area that has a tile index.

Storing these large files and distributing them can quickly become an issue. Thankfully, a free and fantastic .las compressor is available called LASzip which outputs .laz files. LASzip is written by the same software developer as LAStools, so the syntax is similar. In one example, a .las tile is 91MB, but using LASzip the file compresses to 25MB, about a 73% reduction. There isn't a noticeable performance difference working with .laz compared to .las files.

A large number of government agencies, namely USGS, have adopted the .laz compressed open format to store and distribute LiDAR data. This is after a famous commercial GIS software company attempted to take over the .las data format with their own proprietary format which, of course, requires their software to read and write it. The author of LAStools also wrote LASliberator, a program to free LiDAR points from this commercial vendor format.

Point Classification and Extraction

LiDAR files contain metadata in the header, such as projection information. Lasinfo is a quick command that scans a file and reports how many points are in each classified layer. Point classifications are usually performed by the LiDAR vendor and coded into the file for each point. Users can assign these classifications with open source software too, but there are a lot of variables that need to be considered to classify LiDAR points. Different terrain types may require slight adjustments in order to produce a smooth bare earth model.

There are usually at least two classes in a LiDAR point cloud: 1-Unclassified and 2-Ground. Bare earth DEMs are only concerned with ground points, but other classes can be incorporated like:

- 11-Road Surface
- 3-Low Vegetation
- 4-Medium Vegetation

- 5-High Vegetation
- 6-Building
- 9-Water

Simple, fast and efficient command line tools can reproject and extract .las or .laz points for only the desired classifications. LAStools can also convert the files to a simple ASCII X,Y,Z file for input into almost any GIS application. The speed at which this software can slice and dice billions of LiDAR points is absolutely incredible.

LiDAR Processing: Binning and Thinning

Binning is a data process that converts continuous data into discrete data. One example is to assign one value per cell, even if there is more than one data point per cell. The idea for the initial processing of LiDAR data is to figure out what size grid cell is appropriate for the DEM. Too much thinning can create generalized DEMs, while little or no thinning can cause excessive processing time. The trick is to find balance between the two extremes, although one meter resolution has generally been a very good fit from experience.

The process involves binning the input points to discrete data cells through a rasterization process. For example, starting with 10 foot cells, import the LiDAR point data. If the points are closer than 10 feet from another point, then thinning will occur because many points are converted to one cell. Only one value can be assigned to a raster cell. If there are multiple points per cell, we can control which value is contained in the raster cell. For hydrology, we want the minimum value to ensure that small features like ditches and streams are captured with their bottom elevation, not an average from banks. We want to model shallow ponding areas too, so importing the minimum value provides us with the worst case scenario.

Point density can be a useful process to determine the optimal cell size for a LiDAR DEM. There's some trial and error involved, but it doesn't take very long. The goal is to try and get about one cell per point. It's an unlikely outcome because of tree canopy that obstructs the ground, but in open areas, this is a good estimate. We also want to minimize the number of null cells because that means more guesswork for the interpolator when building the DEM. One useful technique to determine point density is to use a statistical sum for each cell.

For example, if using a 10 foot cell DEM and LiDAR points are statistically summed for each cell, let's assume this output:

- 3 cells with no data points
- 50 cells with one point
- 300 cells with two points
- 200 cells with three points
- 100 cells with four points

What does this tell us? Most of the points don't meet the one point per cell goal, so we should probably lower the cell resolution and try again.

Now let's assume we bring the cell resolution down to one meter and import the data points again. Here we find that the numbers shift in our favor:

- 10 cells with no data points
- 500 cells with one point
- 20 cells with two points
- 15 cells with three points
- 5 cells with four points

Selecting an Optimal Raster Cell Size

The best kind of data to create a DEM with is contour lines. What? If you've been paying attention, this should jump out as completely incorrect. What we want is, high-density points with a very good spatial distribution of X and Y locations. The first thing to do is to review the data points and choose an appropriate cell size. Literature suggests that cell sizes can be half the size of the point spacing. While we all love great detail, it's important to remember that just because data supports modeling at a given resolution, it doesn't mean the data should be modeled at that resolution. Or even can be modeled without exhausting system resources.

This balance between what is possible and what is practical is something that just goes with experience. If a number is sought after, one meter resolution data is a good balance between capturing features and performance. Raster data is continuous, but even areas with no data hold a null value, so large raster layers can be enormous resource hogs. Resources may become exhausted trying to build large, detailed DEMs. At worst, after hours and hours of processing, the interpolation program might fail. At best, they can take an enormous amount of time to create.

Big raster data not only takes a long time to create, but then there is the compounding issue of subsequent geospatial functions taking longer to run too.

The primary driver for raster file size is cell size; Okay yes, arguably extent - they're interrelated. While we all can't wait to get detailed DEM results, sometimes the best approach is to use a larger cell size and get faster results at a coarse scale. A cell size in the order of 20 feet or larger is large enough to overcome small details like culverts missing in roads. Different cell resolutions or extents are an excellent use of the GRASS GIS mapsets. Getting results with a coarse resolution is better than waiting days for a crashed software and no output.

Creating a DEM From LiDAR

There are various software tools to create a DEM, but the concept is the same regardless of the software or tool used. Creating a DEM involves taking a bunch of input points, or lines, and creating a continuous surface. It's continuous in that every value interior to the DEM is filled in with a value. Raster interpolation is the process of using known data points to guess the values of unknown cells. There cannot be null cells in the interior of the raster, they all should be populated with a value. Hydrologic flow routing, and other analysis using viewsheds, depend on a continuous surface for the results to make sense.

There are different functions that can interpolate a surface. Some algorithms localize the influence of the data points such as *Inverse Distance Weighting (IDW)*, while others use splines to distribute the point influence over a greater area. Others are more sophisticated with a greater selection of parameters to adjust, such as GRASS GIS's v.surf.rst command, which uses *Regularized Spline with Tension (RST)*. The RST module has some issues with voids over large areas such as lakes and often needs some tuning. Due to the segmented and localized approach of RST, large void areas may have artifacts present in the DEM output. If the voids are lakes, we can beautify them with hydro-flattened features.

Artifacts are represented as sharp and ugly 90 degree angles in the terrain, with edges that do not exist in reality. Other algorithms using a non-localized approach can generalize the surface by performing a best fit to the points over a larger area. Spline interpolation is a great choice for this. Natural neighbor is a really good interpolator with dispersed points for input. Consult manuals in the software for the appropriate type of interpolation to use for specific applications.

Refilter LiDAR Points For More Detail

Up until this point, we assume that LiDAR point clouds are classified by a LiDAR vendor and we just extract the appropriate points, which are mostly ground points. Some situations may require having another look at the LiDAR point cloud to see if any additional detail can be extracted. GRASS GIS has LiDAR modules that can process point clouds into a DEM and DSM. Another open source tool is MCC-LiDAR, a command line open source software that can filter lidar point clouds. If you want to spend some money, LAStools uses some advanced programs that can filter LiDAR point clouds.

There were a few times that a higher detail DEM was required, but one instance consisted of about 20 tiles. The initial machine for Linux and geospatial was maybe eight years old and recycled (upcycled?) for open source GIS. There were billions of points and this required a lot of memory to do the classification, or a customized segmented approach. A custom index was established and looped through to process small areas and join them with the larger ground point collection. This program was written in Python and ran GRASS GIS modules for ten days without fail. The system and programs kept on chugging and in the end there was success. Open source proved to be durable, where commercial GIS software had a hard time making it through the day without a hangup or crash.

Breaklines: A Bad Idea, Mostly

"But my project has these breaklines that the commercial GIS software is pushing, shouldn't I use them?" No, just skip them. LiDAR points are survey points. A breakline is a linear feature that has the same elevations for a line segment. They're like a topographic contour, except that they touch other lines that are not the same elevation. Breaklines are elevation features that are used to create specific features in the DEM or to populate areas with voids. Examples are road edges, lake edges, and stream banks.

Breaklines used to supplement a LiDAR point source are not survey data, they're derived from other methods such as photogrammetry or technician interpretation. Breaklines signal a terrain change, which may be abrupt or subtle - sometimes called hard and soft breaklines respectively. Breaklines are usually provided with a LiDAR deliverable from a vendor. Other features provided may be: classified .las files, DEMs, and topographic contours extracted from the DEM.

The beef with breaklines comes from these buggers' ability to ruin an otherwise perfectly good LiDAR DEM. Breaklines have several applications, but a common one is to hydro-flatten features in a DEM. Hydro-flattening removes LiDAR points and replaces the elevation values with a constant value for closed water bodies, or sloped elevations for features like streams and ditches. Hydro-flattening lakes is scientifically correct and visually appealing.

In one experience, a vendor delivered a LiDAR data package that used break-lines to hydro-flatten a lake. Sounds great until we realize that the vendor assigned the wrong elevation to the lake edge breakline. This several foot error in the breakline elevation created a "lake plateau" when the DEM was created. Lake plateaus are a serious, yet funny (ha-ha) problem, especially for hydrologic modeling because water can't route into the lake. The plateau acts like a wall, so water is routed around it, rather than into it. Consider an even worse problem, where streams become walls that block and divert flow, rather than accept it. Stream walls, how interesting.

Forcing Hydrologic Flow in the Wrong Direction

In another experience, there was a major issue with a ditch that was hydro-flattened using breaklines. Hydro-flattening a ditch involves using lines that parallel the ditch feature to ensure elevation drops occur in a certain direction. In reality, we know that ditches do not have a perfect slope or even a flattened bottom like a concrete swale. Several real-life impacts make perfect ditch geometry impossible, such as vegetative growth, livestock crossings, vehicle crossings and more generally, simple erosion and deposition.

Flow impedance in ditches becomes even more amplified when the surrounding terrain is flat. A flat terrain means there isn't enough elevation change to overcome flow impediments. In reality, the flow becomes interrupted with possible ponding in the channel; aka standing water. This behavior is also applicable to non-perennial streams that have interrupted flow, but since they are natural they tend to flow better in the terrain.

If there is a flat area with a hydro-flattened ditch running through it, the ditch is the determining factor for where the water flows from the flat area. Based on the experience with breaklines and the imaginary walls they create, this sounds like risky business. Why wouldn't one use a nice distribution of survey LiDAR points to model the terrain and let a flooding module report where the water goes? It's certainly more scientific than trying to incorporate other elevation features into the mix and making a judgment about flow in a flat terrain.

In one experience, a scientist made the erroneous assessment that drainage from a potential development area went to a protected watershed. This is nothing to take lightly, altering the drainage for a feature in a protected watershed was not allowed. What were the driving features behind this erroneous conclusion? If you guessed breaklines, then you have definitely been paying attention. During the DEM creation, a technician inserted perfectly formed ditch features into the DEM, forcing flow in the wrong direction.

In this corner, we have breaklines shaping features in a DEM. And in this other corner, we have GIS analysis and an engineering survey. Which one is right? It was interesting, because using the vendor classified data it was unclear which way the area flowed, the area was so flat that in a large storm it probably flows both ways. But when the data was refiltered to greater detail, it was found that the water flowed to the south as the engineering firm concluded. These combined experiences put the usefulness of breaklines into serious question. They already were not being used in terrain modeling and after this experience, they would never be used going forward.

Breaklines in Urban Areas

Accurate breaklines can add value in certain situations. High fidelity breaklines in an urban area can be useful, mostly because the features in a city are features made by people, not nature. Even in nature, breaklines (as polygon) might be useful to hydroflatten a lake in a DEM. There might be areas where LiDAR ground points are obscured by trees or low-lying vegetation. Even if we agree there is a use for breaklines in these scenarios, we're back to the original problem: What if they introduce errors into the DEM?

Sure, the breaklines can be run through an elevation check against the DEM as described later, but since they come from different sources, that's another batch of rectifying and checking that may or not be worth the time for what is gained. High-density LiDAR point data negates the need for breaklines, especially in natural landscapes. Until some new technology emerges, let LiDAR tell the story about the elevations in the terrain. There have already been enhancements for LiDAR penetrating vegetation. The technology just keeps getting better, so be careful about what you or others shove into your LiDAR DEM.

The most valuable lesson learned while exploring breaklines, was to be cautious about using them at all. The next important lesson was to approach any DEM provided with caution; i.e., lake plateaus and stream walls or even just surface filling. Build your own DEM when you can from LiDAR. It's not difficult and this little side trip will be well worth the investment of time. Maybe you can find something cool to do with LiDAR or explore a new idea to share.

Go Forth and Explore

Not only is LiDAR point data the best source data for a DEM, but it is commonly available to download for free. Many government agencies have captured elevation with LiDAR as part of a national flood hazard modeling effort. Data may be available at the federal, state, regional or local level. Be sure to give a good look, it's a resource worth hunting down.

The hope is that this section about LiDAR has inspired others to investigate this incredible resource for GIS. It is an eye-opening experience to see the geospatial products that can be extracted from a LiDAR DEM. The next chapter gets into some of these: potential seepage slope identification, drainage area and drainage network delineations, potential wetlands and topographic depressions. These are just a handful of features easily extracted from a well-crafted DEM. We need you to join the fun and add more capabilities to the list. Give it a go and see what can be discovered. Most of all, have some fun with this amazing technology.

7.5   Garbage In, Garbage Out

The rapid uptake of LiDAR technology had quickly elevated it to expert status in the application of LiDAR point data. The word application is key, because there is no claim of expertise with LiDAR data, but rather the application of the data in GIS technology. This expertise was used to solve a problem the district agency had: LiDAR vendors providing bad DEMs. These are the now infamous lake plateaus and stream walls.

The problem would start with LiDAR vendors providing a LiDAR project package to the agency. The package usually contained at least these datasets: DEMs, LiDAR points, topographic contours, breaklines and .las files. The agency would then provide the LiDAR data package to the contractor or consultant to perform the floodplain modeling. When the floodplain modeling failed because of bad DEM data, the data had to go back to the agency, and then back to the LiDAR vendor to fix the problems. After the LiDAR vendor revised the DEM, the vendor would send it to the district agency, then to the floodplain contractor and hope it's good this time.

It should go without saying, an unbelievable amount of time was lost before the floodplain modeling even began. The agency had a verbose document that was supposed to provide LiDAR vendors with specifications detailing how to provide good LiDAR packages. After reviewing the document, it was apparent that it needed to be rewritten as a usable QA/QC checklist for vendors. All of the previous knowledge and experience about LiDAR would help contribute to this checklist. The verbose and detailed document was decoded and turned it into a concise checklist that vendors, or the agency, could use to rapidly review and check LiDAR data products. It was later told that this checklist was heavily used by vendors to assure complete product delivery.

Breakline Quality Assurance & Quality Control

Building on this successful transformation of document to checklist, a meeting was set up with the agency to discuss point cloud filtering and QA/QC of breaklines. A process was developed to identify bad breaklines, which quite frankly, seemed overly simple and intuitive. In urban areas, the breaklines added some value, but in natural landscapes, they were determined to be without value. As stories about lake plateaus and stream walls made it back to the office, curiosity was more than piqued.

It was a bit surprising that there had not been a solution found for bad DEMs. There was, afterall, an enormous team of engineers, scientists and GIS "Professionals" involved in the project work. The problem was lake plateaus and stream walls, and also the reverse, where features were sunk into the terrain too deep. It wasn't that the LiDAR data was bad, it was the integration of breaklines that ruined the final DEM product. The problem to solve was, identifying breaklines that substantially differed from the LiDAR data. Can you think of a way to test this? Surely if you take a moment to think about it, before the solution is presented, you can solve it too.

Breaklines and LiDAR data came from different data sources. The simplest way to check their vertical differences was to compare the elevations between an interpolated LiDAR DEM and the breakline elevation values. Create the DEM, then using raster algebra, subtract one from the other to obtain the elevation differences. Once a difference raster is obtained, the data can be queried to illustrate features that deviate by X feet or even inches from the surface.

If breaklines were several feet above or below the DEM, they could be flagged and reported as erroneous. This quick check would stop the bad DEM data at the source, rather than finding out the data is bad when it is used as input to the floodplain model. There were plenty of engineers using GIS for floodplain modeling, but no one had any idea how to address this GIS data problem. Someone who is into the data and understands it, is well positioned to examine the problem and use the right tools to get to the right answer. Small, cumulative experiences and differences like these are what separate the GIS professional from GIS users, not some technical certification or test.

Bending Bad Software

All that commentary in the earlier chapters about commercial GIS software was surely just hyperbole, wasn't it? If there is any doubt, this next part should be a hint. This was really stunning. To say comical would be unprofessional, but it was quite funny, yes the ha-ha funny. To say the experience was out of body would be hyperbole, but it was surreal. There wasn't much left to think how ancient and archaic the commercial GIS software had become.

The Flowchart

There was a flowchart that had been developed at the agency and word of this flowchart had made its way back to the office. The claim was that this flowchart covered a whole wall and was the commercial GIS solution to flood-plain modeling. It was claimed to be a technological marvel, a wonder to behold. There was great skepticism at the sound of this, given how ancient commercial GIS was and the problems it had with hydrologic modeling.

After being sent to the agency site to mingle with staff, the grand tour was given. Lo and behold, there was "the flowchart". It was true; this thing covered an entire cubicle wall, and the recollection is that it was over ten feet long. An introduction to the engineer was made, and the pride in the creation of the flowchart seemed uncontainable. Cool, we should all be proud of our work, but there was immense skepticism that engineers and contracted workers from the commercial company were really needed to bend bad software to make it all work.

Commercial GIS software had proven to be a complete joke, was this flowchart a joke too? Commercial GIS flow accumulation modules were ancient and this enormously gross process had to accommodate it. What's that thing called, when an overly complex machine is created to do a very simple task? It's a Rube Goldberg machine, right here on full display, but concealed by commercial GIS software clothing. Nothing against the flowchart designers, they should be awarded; it's not easy to make commercial software do the right thing.

This flowchart was a process that modeled floodplains, but looked like it was instructions to build a rocket. After using open source GIS hydrologic modeling in so many ways, it was curious what this all meant. Was it all necessary? Who knows, but given how poor the commercial GIS software was at hydrologic modeling, the suspicion was that it was far more complex than it needed to be. How can shallow topographic depressions be modeled effectively if the algorithm for flow routing filled them in? Who knows and who really cares.

That was someone else's mess to deal with now that open source GIS software was being used.

Word was that the agency had to contract a programmer from the commercial GIS software company to help "make it all work". This was more money out of the taxpayer's pocket and into corporate profits. This situation began to create a bit of heat under the collar when considering what an open source programmer could have done with this process. This process could have been open and transparent, with collaboration between other agencies to find the best practices. Instead, a commercial GIS company was indirectly looting the taxpayers.

Vendor Lock-In Stranglehold

It was showtime. After converting the agency's document into a workable QA/QC checklist for LiDAR vendors, and solving the "how to identify bad breaklines" problem, it was time to show this information to some influential people in the agency's GIS department. Filtering LiDAR point clouds was a specialized process and open source GIS had some great tools worth sharing. Bet the readers can guess the outcome before we even get there.

The presentation went just fine; all the key points were made. The capabilities of open source GIS were showcased and they acknowledged that this was probably something better than the commercial GIS workflow they were using. They were all too familiar with the same commercial GIS horror stories we have come to know and suffer through. There wasn't any disagreement about the capabilities or quality of the open source software. There was even an admission that this was good stuff.

Can you guess the outcome? It was eventually stated that the agency had purchased too much commercial software to change directions. Everything was already set in place. There it was, the dreaded vendor lock-in. An admission by an enormous agency that they had no alternatives at this point. They couldn't change directions if they wanted to and there was no interest to start using open source or get involved. In truth, it's never too late to start integrating open source software because it can be used in conjunction with all that expensive commercial GIS software too.

Public Dollars, Private Profits

This agency had a lot of licenses. There had to be at least a hundred desktop licenses and the licenses were probably top tier. They also used an enterprise database configuration for data management and offered web services. This is a lot of software to purchase, but worse, how much money was being paid in annual maintenance fees? What would taxpayers think if they knew this information? Would they demand the use of open source software to lower costs? Nah, the propaganda campaign was far too strong to allow any critical thinking of the situation.

Such a shame, really, it was taxpayers money funding the giant wall-sized flowchart. Taxpayers paid for commercial GIS contractors and local contractors to bend code to get around software limitations. It was quite alarming to take this all in. It was everything that was wrong with commercial GIS software, right in one place. There was broken software, vendor lock-in, high costs to the public, hiring special employees and a wall sized flowchart; all of this to accomplish where water floods. Unbelievable.

Commercial GIS software is mostly like a multiplying symbiote. It waits for an unknowing host and offers ease of use in exchange for cash. It starts with one host, then spreads to two, three and more users. Slowly, software pieces are added, such as spatial databases and web mapping. Hundreds to thousands of hours can be spent building scripts and tools around this software. Fast forward 10 years and 100s of licenses are in use with all of the supporting commercial infrastructure. It all starts with one software purchase, so be careful if you are there when this first piece of software is purchased.

Imagining an Open Source Software Solution

Just consider for a moment what an open source developer could have done in this situation with transparent use of code, community involvement and the ability to share and publish results with other agencies. Not to mention the savings of time, money and headaches. Flood model tools could be developed and shared around the globe with all kinds of interesting community involvement. There is absolutely no doubt that something like this would grow and thrive in the open source world with far greater outcomes than paying a commercial software company to help write some code.

Commercial software is able to swoop in and court new customers with the promise of awesome GIS analysis and outputs. They see an opportunity and they pounce on it like a predator getting dinner after a few days starving. They are built for this, marketing new customers, putting on a good show and hyping up GIS technology. They even have a way to shut out alternatives and promote themselves as the "only" GIS. In the US, where they live, users have been duped. For as much as the predator wants to pounce, the prey have been conditioned to be pounced upon.

With open source software, there's no corporate sales person to call in. There's no marketing or propaganda pamphlet to hand organizations. There's no what-cha-ma-call it new thingee to hype up or easy button to press. There's no big corporate presentation to be had. No regional representative or east coast representative to send in. It's just good software that works extremely well with a great community behind it. It just takes the right person with the right ambition to start sharing some ideas. Can you be that person for open source software? We bet you can.

> While it is true that open source software did not have a full-on company selling products and producing marketing, the contrast with commercial offerings led to the creation of the Open Source Geospatial Foundation (OSGeo) in 2006 to help coordinate and, indeed, market many of the open source GIS product you have heard of. This is another story in and of itself.

7.6 Open Source Software Salvation

By the end of this chapter it's 2008. Seven more years have passed and it has proven to be an enormous learning period. Web mapping had been explored as well as a heavy dose of hydrologic modeling. One of the biggest discoveries was LiDAR. This technological marvel brought a new level of capabilities and intrigue.

Open source software, specifically geospatial, has saved another user from despair. It's more than just using open source software because at some point it becomes a mindset. The software, and the community, make the transition very easy for new users. GIS was fun again. The software was cool and so was the community of players. These were really great years to soak up open source GIS and more broadly, open source software on Linux. Maybe this should be called **the turbo years**.

The experience is shared about the agency to provide a real example. An example of how so many people involved in a process can be so ignorant and buy "the best". Too many organizations, like this agency, have been locked into commercial GIS software. If you have the opportunity to get GIS going somewhere, be sure to use what you've learned to influence the best outcome.

The next chapter will illustrate some neat applications of GIS in the real world, instances that saved the client a lot of money and operated in good science. These seven years were great years to grow and learn some cool stuff. The next few years would put all this to the test.

8. Unstoppable Success with Open Source GIS

8.1 Open Source Software Superiority

By now it was 2008. In the previous chapter, seven years of information had been absorbed using open source software and specifically open source GIS software. The paradox was real, the more you know the more you realize you don't know. Open source software, including GIS, was no exception. The applications of GIS are so wide and varied, and go layers deep, that it would take several lifetimes to explore them all.

The value of open source GIS had been proven and there was no slowing down. And the best part, over the hundreds of thousands of acres hydrologically modeled, no one knew any different about the software being used. Remarkable products were coming from open source GIS but since the formats were interchangeable, no one knew it was GRASS GIS until they asked. This was perfect; and how it should be.

8.2 Commercial GIS Software Is Dead

As far as commercial software was concerned, it was dead. There was no desire to use commercial software at all. The computer at work was running Linux for daily tasks. There was a windows machine to remote into, but it was only used for ArcMap work requirements. At this point in the journey, there was a bias to use open source software for everything. Was it perfect? No, but neither was expensive commercial software, and commercial software commanded a huge payout every year in annual fees. Was open source software better? Oh yeah, without a doubt.

From here, the hits kept coming, but unlike the early chapter where hits were something troubling, now the hits were now success stories. It was one after another, after another.

8.3   Modeling Gone Wild

Alright gals and guys, sorry to say we're talking about boring GIS models gone wild, not people. By the time this chapter is over, there would be over 310,000 acres of terrain modeled at one meter cell resolution from LiDAR DEMs. Two 20,000 acre projects and one 10,000 acre project with a one mile buffer for off-site continuity, quickly added to this count. Most of the terrain was modeled in Florida, with troubling flat and depressional landscape. This complex environment became an excellent proving ground for LiDAR and open source GIS software to excel.

Hydrologic Modeling: Process Overview

First we need to note that this information is intended to be a guide, not a technical manual. The process for hydrologic modeling using GRASS GIS is the same for any given area. There are some repetitive parts with slightly adjusted variables, which makes the process perfect for scripting. Bash is a great language to automate GIS processes. This process assumes a DEM exists, preferably from LiDAR. The basic steps are outlined below.

- Set basin thresholds

  - High, medium and low values

- Run r.watershed and r.terraflow
- Extract streams and basins for each threshold

  - High threshold generally to match named large river or stream
  - Medium threshold to match start points in NHD (or field mapped streams)
  - Low threshold to be about 25-50% of medium threshold

- Model Topographic Depressions

  - Extract depressions; subtract surface

- Add culverts
- Add known streams and ditches

  - Helps with flat areas
  - Helps ensure streams and ditches override low confidence areas

- Modify DEM

- Culverts - subtract six feet from DEM
- Streams and ditches, subtract three feet from DEM

• Repeat until all basins and streams are correctly delineated

- Check ponding near roads
- Check sharp angle basins and streams

Basin Threshold

Before running the hydrologic modeling software r.watershed, a basin threshold variable needs to be determined. If you've ever loaded up r.watershed and run it without setting the threshold, this might look familiar, ERROR: Option <basin> requires <threshold>. Wait, there's no magic button to get some outputs? The user has to know something about the data? You sure do, and it's better that way too. That's how real expertise is achieved.

The basin threshold is the minimum number of cells required to trigger a basin delineation. There aren't hard values to use for basin thresholds because the values vary with cell size. A smaller threshold means a smaller drainage area delineation and a larger threshold means a larger drainage area. If the drainage area is smaller than the stated threshold, it is not delineated but kept as part of the larger drainage area that contains it.

On projects with any notable size, it's useful to set three thresholds to arrive at three models of stream and basin datasets. Since the process is in a script, it's just a few more lines of code and the benefits are many. In the 20,000 acre project, we had streams mapped from the field (a story for later) and we used these streams to calibrate the medium threshold. This gave us a known level of specificity in the streams and related drainage areas. If field data isn't available, the *USGS National Hydrography Dataset (NHD)* is a great substitute. Just know that NHD will be slightly less detailed than a field survey.

For the high value, the threshold is dialed in to match major streams or drainage basins. The low value is usually about 25-50% of the medium threshold. The small threshold is intended to reach upgradient into sheet flow areas and model flow patterns that may not be represented by features in the terrain.

Run the Hydrologic Models

After the thresholds have been determined, r.watershed can be run. Another great hydrologic model to run is r.terraflow. It's not essential, but it's useful for at least two reasons. One,:command:r.*terraflow* outputs a hydrologically filled DEM, and two, it also provides another algorithm for flow accumulation to review. We can use the hydrologically filled DEM from r.terraflow to model topographic depressions, discussed later.

After the first pass using r.watershed, two other modules, r.stream.extract and r.basin.extract, should be run. These extract the vector streams and basins from the terrain with given thresholds. Check if the drainage outputs match the desired goals, and if not, then wash, rinse and repeat until they do. After this first run, some adjustments can be made to the thresholds, but it's worth carving in some culverts first because they can impact the size of the drainage delineation areas.

Drainage Enforcement With Culverts

The next step involves slight conditioning of a DEM for hydrologic analysis. Wait, what?! What was all that talk about not altering survey LiDAR data then? Why would this be any different than filling a DEM? Well, there's a few things to consider when performing hydrologic modeling from LiDAR data. One is that LiDAR can't see culverts.

Culverts are pipes that allow water to pass under roads. They're everywhere. If they weren't, our roads would wash out. In the case of hydrologic modeling, the road doesn't wash out, the flow gets redirected along the road in the wrong direction. Our DEM needs to model water passing under roads to ensure that drainage networks and basins are properly delineated. One way to do this is to carve or press the culvert feature into the DEM.

Culverts are features that must be considered when modeling hydrologic flow at high resolution. What's high resolution? In this context, a resolution that incorporates roads. Remember that when importing LiDAR data, we want a minimum value for hydrologic analysis. Let's suppose that a 25-foot cell size is used to model an area. If we choose minimum values on import, and the road is only 15 feet wide, the lowest values on either side of the paved road will be captured for cell values. In this example, the road feature would be eliminated. If we have larger cells, we lose smaller details. Using large cells can work to one's advantage for small scale drainage area modeling, where culverts don't need to be integrated.

To integrate culvert locations, a vector layer for culverts should be developed. All this entails is examining the outputs from r.watershed and looking for sharp angles in the stream or drainage area boundaries. Places should be examined where an extracted stream, or raster flow accumulation, intersects a road and runs parallel. Other areas to investigate are where huge water-filled topographic depressions are against a road. When this happens, it is a good indicator that a culvert probably needs to be inserted.

All we do to create the culvert line is digitize from one flow accumulation cell across the road to another one. If you want to get really fancy, digitize the upstream node first, then the downstream node; this adds a direction to the line. Making a judgment about culvert locations can be tricky and is something that gets easier with experience.

Routing Patches For Flat Areas

Flat areas can be very difficult to determine flow direction. In LiDAR from the 2000s, six inch vertical error was commonplace. Six inches of bias in points can easily influence large, flat areas with shallow ponding. It's also possible that the shallow depression is under canopy, reducing the number of ground points available. Whatever the case, sometimes users need to add a little more known data to the DEM. A small line termed a "routing patch" can be useful to guide these low data areas.

We don't want to alter the DEM from the original LiDAR data, but if there is an anomaly or some data that the user knows the model is missing, they should look for a way to integrate it. If additional information and clues are not available, do not alter the LiDAR DEM. If you have to make a guess, let LiDAR do it. If you have better information, integrate it. This isn't quite the same complaint with breaklines, but it does enter that gray area of altering the DEM. The routing patch is just a line that connects flow accumulation in flat, low data areas, and can be subtracted from the surface at three feet.

Streams and Ditches

The same goes for streams and ditches in flat areas. Streams will have a way of navigating to the lowest terrain and are mostly not an issue. Depending on conditions, it might be necessary to add some connections. Ditches are a little different because they can have odd impacts on hydrologic flow and drain flat areas. It may not be clear what benefit the ditches provide, but if you can see them on an aerial photograph, then map them as vector features. In some instances it's helpful to subtract them three feet from the surface if the LiDAR is sparse and the ditches seem relatively new.

Culverts, ditches and routing patches can help guide the flow accumulation, and they are useful vector layers to have for the project. If there are field teams involved in the work, they can help create the initial culvert inventory or field verify the culverts created in GIS. The routing patches and ditches are mostly useful in flat terrain with sparse data, but the culverts are definitely a feature of interest in high resolution hydrologic modeling applications.

Topographic Depressions

A topographic depression can be a surface depression that fills up and flows water, or it can be a large bowl area where the water has to drain as ground-water only. Retention ponds are another example of a feature that drains internally. Internally drained features can be inputted to r.watershed so that surface water flow is modeled to the lowest part of the topographic depression.

The shallow ponding areas can be useful for civil engineering applications. As pipes and culverts are designed, they are designed for storm events that are related to a volume of water. If all of the water has to go through a pipe, the pipe has to account for all of this volume or roads can wash out. There were several times engineers needed GIS modeling to estimate how much water would be held up in shallow depressions, because water held in shallow depressions doesn't need to be accounted for in the pipe design. So, even though we're building depressions for r.watershed, it should be noted that there are other use cases for modeling shallow depressions.

How do we know when to include topographic depressions as input for the r.watershed command? It's back to that experience thing unfortunately, but there are some tips that can be useful in trying to figure this out. As we mentioned, one output from r.terraflow is a hdyrologically filled DEM. If we use map algebra to subtract the DEM from the water-filled DEM, we end up with a raster layer of water depths for depressions. Initially, this result has the potential to cover most of the study area because even a value of 0.1 feet, or smaller, will be included just by way of the math.

Once we have our topographic depressions modeled, we can then filter the depressions to something meaningful. We can look at the depth and area for each of these, and begin to key in on ones that are holding a significant amount of water. The first obvious filter is by depth, so here we can get rid of anything that is less than one foot in depth. Adjust the values as you see fit, this is just a guide. After the depth is filtered, we can filter by area too.

Maybe in this case we only keep areas that are greater than 20 square feet. Now we are left with significant areas that store over a foot of water with a surface area greater than 20 square feet. Remember that r.watershed is looking for areas that are internally drained, so keep this in mind when filtering the topographic depressions. The great thing about wrapping this in a script is that only a few variables have to be tweaked each time.

Wash, Rinse, and Repeat

After we have run r.watershed with some initial thresholds for high, medium and low, we can look to adjust those values on a subsequent run. Also, we definitely want to modify the DEM to include culvert data. Known depressions can be used as input to r.watershed as well for internally drained areas. Rerun the script and keep adding culverts until every delineation and water-filled topographic depression is satisfactory.

The iterative process sounds like a great machine learning exercise. There is a pattern to inserting culverts, mostly associated with flow that takes a sharp turn against a road. When flow accumulation makes a sharp turn against the road, the learning algorithm should continue the flow through the road to the next accumulation cell. There also seems to be machine learning calling out to solve the topographic depression parameters. There's no claim to have all the answers, but the described iterative process seems like something a computer can learn. Maybe you can develop the plugin or extension that adds this capability? Bet you can!

Process Recap

So to recap, this process can go in a Bash script, or Python, it's open source so there's flexibility. First, we estimate three basin threshold values and then run r.watershed. Also run r.terraflow for a hydrologically filled DEM and alternative routing algorithm to review. Extract three pairs of drainage network and drainage area vector layers, using high, medium and low thresholds. Create a topographic depression layer by subtracting the DEM from the water-filled surface. Filter this layer by depth and area. Next to roads, look for water-filled depressions and sharp angles in streams or basin lines. Create culverts, routing patches and ditches as vector layers to integrate into the DEM. Keep re-running the model with updated features, until all drainage areas and stream features look satisfactory.

At the end of the process, we get some cool stuff: three vector flow accumulation networks (streams), three corresponding drainage areas and modeled topographic depressions. The culverts, routing patches and ditches are developed by the user and also have value for the project.

8.4   Modeling Potential Stream Locations

The Case of the Missing Streams

For this next story, it is curious why the problem existed in the first place. Maybe the task was to delineate streams from an aerial photograph. Maybe it was from checking a short list of probable locations in the field. Maybe no one knew about hydrologic modeling. Way too many maybes, but if good GIS software was used in the beginning, with the right person, there likely wouldn't have been this problem to solve. Thankfully, it was now our problem.

The project area was about 20,000 acres. There was a mapped stream inventory provided by another company and our task was to assess each stream for pre-development conditions. A stream ecologist had to visit each stream and record the characteristics. Visiting where streams might be; interesting, this sounds like a GIS problem to solve.

It was a common occurrence to be called upon to solve interesting client problems. Being approached with the, "I have a conundrum" phrase brought great joy because it was for certain there was a complex task involved requiring critical thinking and curious applications of GIS software. In this instance, the situation was not good. The client's permitting, and thus development, of the 20,000 acre site was put on pause. This had potentially catastrophic consequences due to the size of the operation.

As the story goes, there was a state agency reviewing jurisdictional wetland delineations in the field. During their investigation of the 20,000 acres, two streams were walked over that were not mapped in GIS. As a regulatory agency of wetlands and streams, they were obviously concerned. One stream might be missed, but to witness two streams before lunchtime was not acceptable. Finding two streams in a short time reduced their confidence in the mapped stream inventory. They said they would not resume the wetland inspection until all of the streams were mapped.

There was an unmistakable panic in the air due to how long this could take. There was no room for error given the circumstances arriving at this point. This was 20,000 acres in some of Florida's most unforgiving landscape, coupled with the complexity of interrupted streams in flat terrain. There was no room for error given the circumstances arriving at this point. Sounded like a fun problem to solve.

Stream Prediction Model

Remember that the streams had previously been mapped by another firm. How the streams were mapped was unknown, but it was obvious there was missing data. As the problem was described, thoughts of flow accumulation and LiDAR data danced with great vigor and delight. One of the main characteristics in modeling flow accumulation, and thus streams, is a minimum threshold value that creates a modeled stream line. In this case, we were just looking for potential areas where streams might exist. The model would guide field teams in an intelligent way to the areas where water accumulates. Then they could make the determination.

Using a very low threshold in conjunction with field delineated wetlands, the predicted stream locations looked very reasonable. We used jurisdictional field mapped wetlands as another indicator, and together the data was an excellent guide for the stream ecologist and team. Not all of the streams found can be attributed to the modeling, however, the effort yielded a 25% increase in mapped streams, or 66 stream segments added to the inventory.

The numbers alone are impressive, but the real benefit came from using a scientific approach to predict stream locations for field investigation. This assurance to the agency was possibly more important than the number of streams added to the inventory. It wouldn't be the last time the agency would be assured by best practices and data. The stream ecologist, a professional of many years, stated that the stream models were excellent. He'd recommend using this model for all stream projects going forward.

It should go without saying, this success story was a major accomplishment. Just doing a task is one thing, but having agency approval is something entirely different. It's like having peer scientific review. As a public agency, they are held to much scrutiny and inquiry, and being able to give them reliable data was invaluable to the process.

8.5   Modeling Stream Banks

Does the gift of LiDAR ever stop giving? Only if you stop thinking about the various ways to use it. This really was a neat application that again prompted agency approval of a new method and approach to mapping stream banks. The requirement wasn't explicitly to map stream banks, but to map other surface waters. These were non-wetland, water features that the state had jurisdiction over. Streams were one of these other surface waters.

The stream shown in the illustration has been ditched for agriculture, further accentuating the feature. This area was chosen as an example because of the clear features visible in the open terrain, with contrast to the rest of the stream feature hiding under canopy. There is a common myth that LiDAR doesn't work under tree canopy. Even back in 2007, this data still did a remarkable job mapping features not able to be seen in photography. Can you think of a way to delineate stream banks before it is revealed? The bets have been placed that you can.

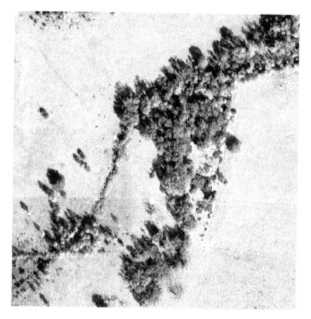

Figure 8.1: Stream Hiding Under Tree Canopy

Always Look to Improve

They say if it ain't broke, don't fix it. It's true, but don't we like to tinker and enhance things? How does anything ever get better if we don't break it? Sure, we don't want to break production to try something new, but we should always be looking to make processes better or improve a system as technology expands. It's foolish to let the technology get ahead of you and play catch up, rather than be a part of the technology change.

As time goes on and technology improves with detail, so do the regulations and requests to obtain a permit. Before LiDAR technology was available, there was no agency request, or practical way, to map the limits of a stream bank.

By the time a traditional survey crew would have finished the project, nature would have already found a way to change the boundaries. Well, not quite, but you get the point.

Regulations now required all *Other Surface Waters (OSWs)* to be mapped with the limits delineated by GPS points. There were about 13 miles of stream banks to delineate and the idea of using GPS to delineate them was nauseating. This was another time-sensitive situation where technology was needed to rapidly meet a deadline and avoid any delays in the development schedule. Like the other crisis situation, this one also had the potential to significantly delay the schedule.

Stream Feature Mapped by GPS

As dire as this situation was, it can't be recalled how it originated. There isn't a recollection of an "I have a conundrum" moment. It's likely this was just investigated out of curiosity, "Could GIS be used to model a stream bank?" An intriguing question, one that couldn't wait to be explored.

Among the first streams mapped with GPS was a stream in an open field. There was no overhead canopy to obscure the GPS signal. The field team used a sub-meter GPS to collect points along the stream bank, mapping significant bends and angles in the stream banks. The GPS points were used to create jurisdictional limits as GIS polygons. As excellent as the fieldwork was, the nature of mapping stream banks with GPS points still presented the problem of over simplifying the bank limits.

Can You Take a Swing?

Time to put on the thinking cap and see if you can figure this one out. What do you notice in the shaded DEM illustration above? What could you do to identify a stream bank? What happens on the surface where the stream bank is located? Are there any parameters or characteristics that we might be able to look at and key in on?

There is probably a more sophisticated approach to this, but in simplicity, slope was examined for this application. The slope changes drastically at the banks and can be queried out in a meaningful way using a detailed DEM. When these solutions are presented to other GIS analysts, the reply is usually, "Whoa, cool, why didn't I think of that?" Not sure really. These aren't complex solutions, but what they do require, maybe more than anything, is an awareness of the available functions in GIS and the connections to reality. Remember all that detail we went into about rasters and cell size? If the goal is to model

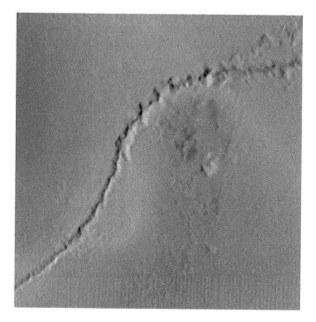

Figure 8.2: Stream Revealed Under Tree Canopy

stream banks, then be sure to use a small enough cell size that can capture the features of the stream, such as the bank.

Feature Extraction

This is fun stuff: finding hidden features in a DEM, then revealing the obvious. At least it's fun for some people, if not, maybe another section or application will be interesting. Sure, these functions can be put behind a button once the process is figured out, but it's the process that is important, not the button. Keep that in mind when there are easy buttons in GIS software; commercial software is infamous for charging extra for buttons they deem special. It's applications like these that get wrapped into an extension and sold to users.

Today, we can get 8+ pulses of LiDAR beams per meter, probably significantly more by the time this book is published. In the early and mid 2000s, when most of this work was done, the density of LiDAR data was roughly one pulse per meter. Even with about one meter post spacing, we can do some neat things such as *feature extraction*. Feature extraction is a vague term, but here it means running functions on a DEM and then converting the raster features to a vector format. For example, since water flows downhill (be ready for everyone to state this obviousity to you), this phenomena can be modeled by connecting the lowest spots in the terrain. A few examples of features that can be extracted from the terrain are: stream banks, ditches, streams, roads, wetlands as topographic depressions, drainage networks, drainage areas and seepage slopes.

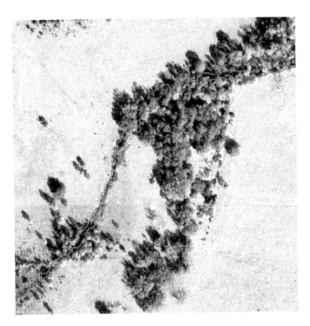

Figure 8.3: Stream Bank Model on Aerial

LiDAR-Based Methodology

The approach using LiDAR relied heavily upon slope changes in the terrain. There were obviously other areas in the project site that had a similar slope that were not stream banks. Since we had already mapped all the streams with high confidence, we knew where to look and focus the efforts. In typical fashion, classified ground points from the LiDAR vendor were extracted from the point cloud. Then the points were used to create the DEM.

Once the DEM was created, slope could be determined for each cell as a raster layer. Different values were used for extraction, as sometimes the stream was obscured by dense vegetation which would change the slope parameter. Be-

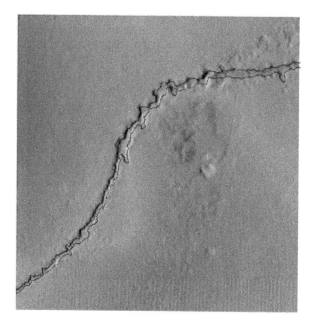

Figure 8.4: Stream Bank Model on Shaded DEM

fore converting linear raster features to a vector line type, it is often necessary
to thin the raster cells.

This modeling technique proved effective for about 90% of the stream banks.
In fact, it proved to be so accurate that the agency dropped their requirements
for GPS points to define stream banks. The LiDAR-based approach proved
equal or better to previous technology. It was the first LiDAR-based method-
ology approved by the state agency that met requirements for mapping stream
bank limits as other surface waters.

Shortcomings of Methodology

It wasn't perfect, but it was beyond a doubt better than the manual alternative
with GPS. It should seem obvious that the banks can only be well defined
if there is an adequate number of points to model the feature. Fewer points
available as input to the DEM means that there is a reduced ability to capture
accurate, sharp changes in the elevation. Tree and shrub canopy can greatly
impact the ability of LiDAR scanners to obtain high density ground points.
In these dense vegetation areas, GPS teams were required to supplement the

LiDAR data with GPS points because the LiDAR points were far too sparse to give reasonable stream bank limits. The GPS teams didn't capture elevation values, they mapped the stream bank limit.

In the example above, it's interesting to note that under heavy canopy (see northern segment) the stream bank appears to deteriorate and widen. Stream banks under canopy areas to the north appear wider than the open field area in the center of the image. The bank may or may not widen under the canopy, however, experience says that the bank is wider in the areas with canopy because there are fewer points available to define a crisp feature. The remarkable thing illustrated here is that the feature is exposed under canopy, dispelling the myth that LiDAR cannot penetrate canopy cover.

Solutions For Enhanced Delineations

GPS teams are a great resource in this effort for two reasons. The first is they can spot check the accuracy of the LiDAR derived stream bank delineations, particularly areas under vegetation. The second thing they can do is collect data in areas where the LiDAR data is sparse.

Another tested approach, which requires in-depth knowledge about LiDAR, is to filter the point cloud to bare earth points. Filtering the point cloud to ground points often results in a highly variable rough surface, rather than a generalized smooth surface. In some instances, this more detailed DEM can help the software extract slopes at the stream bank locations that were generalized with vendor smoothed filtering.

8.6 Water Table Modeling

The water table is also known as the surficial aquifer (thanks hydrogeology professor of many years ago). In Florida, many areas have a water table that is close to the surface. This high water table intercepts topographic depressions to form wetlands. It's relatively easy to model a water table when there are wells that can provide the water table data. Take a bunch of points with known values, interpolate to fill null values, we know the game by now.

When empirical data is not available, which is more often than not, there is another source of information that can be used. The *USDA Natural Resources Conservation Service (NRCS)* soil data is very comprehensive. There are many related tables that can be linked and used for different applications. Buried in these tables is a great little attribute that provides depth to water. These values can be linked to map unit polygons.

Now we have polygons with a depth-to water value, and even a depth to water for each season, wet and dry. The soil polygons with depth to water attributes can be converted to a raster layer maintaining these values. Once in raster format, the depth to water values can be subtracted from the LiDAR DEM surface, one for each season. While the math and concept is simple,[25] the approach to build this layer requires a few other touches. Before we do the math, we need to prepare the water table model.

The land surface is highly detailed and can have a lot of local variation. The water table is saturated water in the ground, near the surface. A water table surface generally mimics topography. One main difference is that it doesn't have all the little elevation changes that a land surface model has. Its characteristics are smooth gradual changes over space, not abrupt ones (save special features like seepage slopes). How can the water table feature be modeled?

After much trial and error, the best approach was to use a cell size that is large enough to capture the smallest map unit polygon feature. Another great use for GRASS GIS mapsets, a different cell size for modeling the water table. If the water table uses a small cell size like the DEM, then the water table will be too rough. Not only does the water table surface need to be smooth, but the abrupt changes between adjacent "depth to water" areas will need to be smooth too. The water table is a smooth feature and our data needs to represent this characteristic.

A great tool to smooth rough rasters and feature edges is r.neighbors. This function is a moving window of cells. The user can set the number of cells, but if we assume 9 and a mean function, the center cell will be the average of the other eight. So a moving window must always be odd. A small window is good in this case because we don't want to influence cells much beyond the abrupt edge changes, we just want to blend the edges. Smoothing operations like this must be done dozens of times. Got that scripting handy?

After converting the polygons to a raster format with the "depth to water" value, we can subtract the depth to water from the surface topography. Smooth the water table surface with r.neighbors and, voila! One water table surface. Water table elevations are usually best illustrated as contour lines. Contour lines are also a great way to represent raster surfaces to check for any anomalies. They should be very generalized, but mimic topography.

One cool thing you can do is display the water tables and land surface elevations using a 3D viewer called NVIZ in GRASS GIS. This is similar to viewing rasters in 2D space, but using 3D adds an interesting perspective. An example of this is shown in a subsequent chapter with lake volume modeling. When

---

[25]water_table_elevation = lidar_dem - depth_to_water

the two models, the land surface and water table surface, are viewed in 3D space, the intersection highlights areas on the map that have surface water exposed.

8.7 Modeling Potential Seepage Slopes

Seepage slopes are neat environmental features where the water table intersects an abrupt topography change, usually into a wetland. Seepage slopes are considered environmentally sensitive areas, especially when the area is going to be developed. By now we have gone through some tools that can be used for identifying potential seepage slope areas. One of the simplest approaches is simply examining steep slopes with wide limits, along wetland edges. In the project we were working on, we had wetlands delineated in the field. Otherwise the *National Wetland Inventory (NWI)* is an excellent datasource if field verified data is not available. We can also model wetland areas by identifying shallow topographic depressions in the landscape.

The project already had some seepage slopes identified, so we cheated a little and used those slope values to find other similar locations. Without this data though, it's not difficult to start extracting some steep slopes against wetlands and start looking for the outliers. With a little practice, these features can be added to a field map and field verified along with other verification efforts. The idea is to identify potential areas that need field verified, just like the stream prediction effort.

8.8 Stage and Storage Modeling

Remember those shallow topographic depressions in the terrain? The ones captured by LiDAR and too shallow to be modeled as their own internal drainage area? Ah, who cares, it's hard to remember what happened five minutes ago in some of this GIS goo. These shallow ponding areas may occur when it rains, or they may be present when the water table intersects a depression like a wetland. Either way, the modeling process for these shallow depressional features is very accurate.

Topographic depressions can be modeled quite easily in GRASS GIS. There is a module called r.lake that will fill a depression to a specified elevation from a starting seed point. The depression can be filled to an elevation where the water overflows, sometimes called a pop-off elevation. Keep flooding the depression iteratively, until the pop-off value is found, then dial it back a bit. r.lake will also output the maximum depth, surface area and volume of the water storage. Sometimes these models are called stage-storage because the areas are filled up to a certain stage level with an associated volume of storage.

8.9   Flood Depth Modeling in Confined Corridors

Now we can take these two previous model examples and apply them to something slightly different, but close enough to build on the ideas. That's what a lot of this is using GIS and people don't seem to get it. It's years and layers of learning, experiencing and applying. Trial and error. Failures and successes. GIS users who use the software surficially or just push buttons to get results rarely excel into these areas. Always be curious and take the time to learn about what's going on. These layers build a tower of experience and the view from the top is quite amazing.

This example was another scientific model that impressed engineers. It's not to brag or boast, it's to state the facts and show that GIS data modeling can compete on level with engineering practices; sometimes better. Remember the basin delineations improved by GIS modeling? Here's another one.

Transmission easement corridors can vary from 75 to 150 feet wide. Oftentimes transmission lines run through remote areas and have limited access locations. In remote areas, maintenance crews may need to travel within these linear easements. Sometimes they would encounter water-filled depressions in the corridor that could be dangerously deep. How could they be made aware of these dangerous areas?

The project engineers knew that GIS and LiDAR had done some incredible things in the past, so the problem was brought forth. There were a few variations to this process over time, but initially the answer was yes, we can model that. The confidence level was so high at this point, it felt like nothing that could stop this train.

Transmission corridors are linear with angles and bends in them. This makes for some special work in creating the DEMs, because the raster data is confined to a linear corridor. Linear corridors can span large areas and this means a large region processing extent is required. Remember, all those null values still have to be accounted for, so for these large areas you may have to patch several regions (mapsets) together to create a composite DEM in this case. Some DEM algorithms ignore raster masks, increasing the time to create a DEM.

We used soil data to get depths to water in the past but there was a new request. The project engineers wanted to use floodplain data, which would make the modeling more accurate. There were updated, preliminary floodplain transects available for this project. The transects could be used to create a water table and on the first pass they did a great job.

One immediate issue was that there were areas in the terrain that lacked floodplain data. After working with so many hydrologic features, looking at aerial photographs and using DEMs, it wasn't hard to estimate the missing values. The modeling results were put into a tiled mapset and the water depths were illustrated with a nice blue color ramp. It was a hit. The major power company said they laminated the map book and provided it to field staff. Then they gave us several more to do. There were even some corridors that had no floodplain data and they were just estimated based on experience. Those too were a hit. Like stated earlier, the hits kept coming but this time in a most welcoming way.

8.10 Viewshed Analysis aka Line-of-Sight Visibility

It's called viewshed analysis, but it involves a great deal of modeling. There were a few interesting opportunities to model line-of-sight visibility, also known as viewshed analysis. Viewshed analysis uses a point at a location with a given height and scans the terrain to determine which other locations are visible. The ground elevation plays a role in visibility as well as elevated features like buildings and trees. It is possible to use a bare earth DEM for this process, but it would be the most conservative of approaches.

A Digital Surface Model (DSM) represents all of the surface features. In a LiDAR point cloud, there are usually at least two classes: 1-Unknown and 2-Ground. If we want all of the non-ground points, we select class one. Note that these are really unknown points. They could have errors in them or even capture birds in flight. Outlier points should be removed and the DSM elevations should be reviewed for the maximum realistic value in the DSM features because they are unknown.

If there are ground, noise, water and unknown classes, be sure to pick only relevant classes. Classes like noise should be discarded. As an aside, class nine for water is a tricky one. Water can have an unreliable return. It can absorb some of the beam or otherwise give unreliable results. In practice, these points can still be useful and should be reviewed when considered for use in a DEM.

When modeling surface features, it might seem intuitive to use interpolation. Maybe it makes sense in other situations, but when modeling at one meter resolution, features are already exaggerated to the size of a cell. For example, if one inch of the branch is captured by the LiDAR scan, a full three foot cell will be used to represent the one inch section. This modeling will exaggerate the size of a feature by up to almost the size of the cell. Maybe further based on diagonal measurements. It was observed that interpolation would further exaggerate these features if there were null values between the DSM features. Some interpolators might even alter the ground points upward as an undesired side-effect.

A reliable method was found with the beloved r.in.xyz function. This function just imports values to a cell without interpolation. If we first create a bare earth DEM from LiDAR ground points, then we can replace DEM values with DSM values imported from r.in.xyz. This product proved to be an excellent model to perform viewshed analysis with.

When performing the viewshed analysis, there's some other things to consider. Let's suppose one scenario has a cell tower being erected and the question is, "Where can this tower be seen in the community?" Using the DSM we built above, this is a straightforward use of the viewshed tool that will light up a map with areas that the tower can "see". It pays to read about other people's approaches, because one user said after running the analysis, they then removed the visible areas that were greater than six feet. Why? Because the tower can see the top of the tree canopy, but if you are under the canopy in the tree area, you probably have reduced visibility of the tower. If you are in an area where the vegetation is less than six feet, it will not be an obstruction. Nice catch.

8.11   Volumetric Modeling

There is a misnomer with the term "3D modeling" in GIS. Three dimensional modeling certainly implies three dimensions, but let's think about what this means in GIS terms. In surface modeling, we are concerned with only that, the surface of something: earth, water tables, canopy, etc. But what if you had to model the extent of an underground contaminate? What data model could be used since vector and raster data are 2D and cannot model true 3D objects? "Volumetric pixels", the crowd cheered.

Cut and fill operations in GIS don't require a true 3D model. By true 3D model, we mean the ability to map a single location with two Z values. If we picture a bowl in the landscape and we want to know how much data is required to fill it, this can be calculated with a 2D raster layer. We know the X and Y cell resolution (length and width) and we know the Z fill value (height). The basic geometry volume equation is length multiplied by width multiplied by height. We are modeling the difference between two surfaces and that difference can be summarized as volume. GIS models nature and it's rare that natural phenomena need true 3D modeling; cliffs would be an exception.

Of course there was a natural curiosity to explore 3D modeling with voxels. It was a bit troublesome, but in the end, a volumetric interpolator in GRASS GIS was used to model the extent of an underground contaminate. Data points from wells were used as input to the 3D interpolation. It brought back memories of the groundwater modeling at the first environmental company with ARC/INFO, but this stuff was much more sophisticated. There wasn't even a volumetric interpolator in the commercial software.

The data was exported from GRASS GIS and imported into Paraview, an open source scientific visualization tool. This tool could slice through the plume and illustrate the concentrations from within. It was really remarkable technology and worked extremely well with open source software.

So the next time someone tells you they need 3D modeling, just be kind and smile back, "No problem, can you explain more?" Most likely, the answer will resemble something that can be done using rasters and volume calculations. Hey, that's our little secret, no need to be picky about the details of language, really. If they call it 3D modeling, just go with it. After all, 2.5 rounds up to 3 anyway.

8.12   Site Suitability Analysis

Site suitability is a type of model. It is a process that ranks areas, cells in a raster layer, based on how suitable they are for the desired request. Site suitability is a common use of GIS analysis and GRASS GIS has all the tools to accomplish this. Setup the raster layers, rescale them from least suitable to most suitable (0-1), add layers together, consider weighting, remove unusable areas, then shade results based on suitability score. As an example, let's look at a solar power plant suitability analysis. We'll need: open land greater than 200 acres, wetland setback areas, and proximity considerations to electric transmission lines and residential areas.

First, we identify suitable open land use areas which can be done by assigning a one to agriculture or pasture polygons, a five to shrubs and brush and a zero to forested areas. Note that a suitability score of zero still considers an area and does not remove it from the analysis.

Then we look at wetlands and their setbacks, areas that cannot be developed. The wetland and the buffer should be assigned null, not a suitability score. We don't assign them null because it's unknown. We assign them null because when they are used in the map algebra, the areas will be discarded and not used as part of the suitability analysis.

A pretty cool function in GRASS GIS is `r.grow.distance`. This takes a feature and assigns a distance value for every null cell around the feature. So, for our example, locations are more suitable if they are closer to a transmission line, and less suitable the farther away they are. The numbers should be rescaled from zero to max distance, to one to zero. A suitability score of one indicates closest to the transmission line and a zero indicates the furthest point away. The distance buffer lies at the heart of site suitability analysis.

We can use the same approach with residential homes, except the suitability scores are reversed. It was preferable to be near a transmission line, but it's not preferable to develop a solar power plant next to a residential area. The further away, the better, so for proximity to residential areas, we set zero to max distance and rescale the data from zero to one.

When the suitability scores are combined, electric transmission line proximity + residential home proximity + land cover + setback buffers, an overall suitability score is obtained. Weights can be assigned to the layers in the above equation. If it's more important to consider electric transmission line proximity than being away from residential areas, the equation can accommodate the consideration. Once all the data is combined and rescaled zero to one, areas can be filtered to show the top 25% most suitable locations (or whatever threshold is applicable).

8.13 Enterprise Capabilities With PostgreSQL & PostGIS

What Is PostgreSQL & PostGIS?

One word: awesome. Just like most open source GIS software, this database software is awesome. These two buggers were always lurking in the background of the journey. References from GRASS GIS software and mailing lists made it perfectly clear that this database technology was to be used when limits of desktop or file-based GIS become cumbersome or restrictive.

PostgreSQL is an amazing *Relational Database Management System (RDBMS)*. *PostGIS* is referred to as a "spatial extender". PostGIS stores geometry data types in tables. It can perform functions with the tables (GIS layers) using *SQL* statements. While PostgreSQL has its own geometry format that can be used, PostGIS is commonly used because of its widespread acceptance, spatial functions and data types. For GIS applications, they are usually installed together.

Can't really recall when the LiDAR in PostgreSQL occured, but it might have been after PostgreSQL was up and operational for the GIS data library. Either way, we'll sandwich this in here even though the dates might go a bit out of the range specified for this part of the journey.

PostgreSQL was always of interest. A few books had been purchased and read, and the RDBMS was installed on the local Linux system. It was used a few times to process big data that had encountered limitations in Microsoft's Access database. In all likelihood, something about LiDAR and PostgreSQL was encountered on a mailing list. There was this crazy idea that LiDAR could be stored and accessed in a RDBMS. Wow!

LiDAR Data Management

Whew. The IT gals are not going to like us on this one because LiDAR data can take-up a lot of server space, even with LASzip. A common way to store LiDAR data is to use a file-based management approach. The files are usually stored by LiDAR flight mission or project in a directory. The process to fetch the LiDAR tiles was repetitive and always the same.

First, buffer the area of interest one mile. This would create connectivity off-site and help model drainage continuity. Then, select LiDAR index polygons that overlap the one mile buffer. Fetch the tiles from the Windows network server, and then process them into a DEM in GRASS GIS. It's a straightforward process and after doing it a few times it became clear that there must be a better way. Experience has shown that if the thought is encountered "there's got to be a better way", there is.

That alternative way was found in PostgreSQL and PostGIS using a LiDAR extension. Not really sure if the same one was used at the time, but pgpoint-cloud sounds familiar. This amazing extension allows LiDAR data to be stored in PostgreSQL, queried and displayed in QGIS. Testing the capabilities, a LiDAR tile was loaded into PostgreSQL and then symbolized by elevation and displayed in QGIS. Using a RDBMS would be an excellent opportunity to optimize LiDAR data retrieval for GIS. Queries could select the most recent data available and directly import them into the GRASS GIS pipeline to create a DEM. We never got that far, but maybe you will.

Mobile Data Collection

The mobile "phone", or more appropriately, mobile device, is an amazing piece of technology; the capabilities far exceed that of making a phone call. Literally everyone is using a touchscreen device today, ranging from toddlers to grandma. The relatively simple and limited interface makes them easy to use. This ease of use has woven them into our everyday life and society. Mobile devices, phones and tablets, have many built-in features such as location by GPS, maps, photo camera, video camera, compass, virtual keyboards and talk-to-text. These built-in features allow them to be used quite naturally in the scientific community to collect data.

Mobile data collection systems are designed to make the phone or tablet an easy-to-use rapid data collector and also provide a database system to aggregate and store the data. Mobile devices can be used to enter text or numeric information, record locations with GPS and record media such as photos or video. This system is easily scalable to match the project demand, by allowing one to hundreds of people to collect and submit field data using a standard-ized form for input. In addition to the collection and server software, there is usually a component to build the forms that are used in the collection software for end users.

In our situation, choosing open source technology over commercial software would provide stakeholders an initial purchase savings of $20,000, followed by an annual savings of $12,500. That's a savings of $70,000 over a five year cost of ownership. The mobile data collection system can be run with existing staff capabilities and with minimal training. Two commercial software sys-tems were investigated for comparison: *ArcCollector* and *Fulcrum*. Assuming 20 users for the scenario, every year ArcCollector costs $10,000 (plus storage fees) and Fulcrum costs $6,000 annually. If ODK (discussed below) is too much to maintain, Fulcrum is a great alternative.

Mobile Data Collection Profits

The large commercial GIS company was in the mix again. How are they prof-iting this time? Any time mobile collection data passed through their sys-tem, credits had to be consumed, although they claimed it was pennies. Talk about vendor lock-in, there was no way to collect data and just import it into the desktop GIS. No, the data had to go through their online service where it would consume credits, a storage penalty. Since no one ever leaves gobs of GIS data lying around anywhere (note heavy sarcasm), this would surely be a rare event to get hit with cumulative credits for data passing through or residing on the system.

Exporting data required exporting to a geodatabase, stored in their online sys-tem and then downloaded. Oh sure, they gave you some credits "for free" with maintenance, but we already discussed how none of this stuff is free if you are paying an annual fee of tens of thousands of dollars. Vendor lock-in had moved to the cloud; of course, why not, the cloud is magical and they use magical credits to charge for it.

Open Data Kit (ODK)

Another adventure that used PostgreSQL was the *Open Data Kit (ODK)*. This was really cool technology. At this time, in 2017, the organization had not adopted any online web applications or mobile data collection. There was field work based heavily on GPS data collection, which was working quite well, so don't mess with it. But what about improvements to the process and efficiency? Not to mention other benefits like standardized data collection and these handy things we had now, called phones? Note that this was done back in 2017 and the ODK software was undergoing a transition from university development into an open source project. Surely some things have changed, so please keep this in mind.

ODK is just like an online service that provides data collection on mobile devices. There are five components that made up the ODK data collection system: mobile user collection software, database server software, web server, Tomcat web server, and the ODK software itself. Of course we can't forget the underlying foundation that makes this all possible: Linux.

There was nothing really exciting to report about the adventure, everything worked as advertised. Well, it was exciting to be running our own online data collection system. A test form was created in Excel and then uploaded to ODK. The user could download the app from the mobile store, then download the form from the PostgreSQL server, then start collecting data.

Installing the ODK software, Apache and Tomcat were without issue. The test subject went in the field and collected some data. When they came back into the network, the data was automatically uploaded into the PostgreSQL database system. It worked extremely well, an open source solution to mobile data collection. Put this PostgreSQL database on a Linux cloud server and done. We never got that far in the adventure, but if you do, please make it known.

One important thing learned in this adventure was to increase accuracy of the mobile device. Mobile GPS is around 15 feet accuracy, but this can be increased with an external bluetooth GPS receiver. They can clip on the shoulder and are high enough to get a good signal. This hardware addition creates an opportunity to collect some really accurate GPS data. Of course this great system had to be reviewed at before writing this, and wouldn't you know it, there is a QGIS plugin for ODK.

8.14   Unquestionable Success Using Open Source GIS

It was now 2015 and a lot had happened in this section of the journey, a lot of great things. For the last eight years, all analysis and modeling work has been done on Linux using open source software. This was a remarkable accomplishment given how extensive and pervasive the modeling had become in the organization. ArcMap was used for cartography due to organizational standards, as well as specific client deliverables that required an ESRI format, specifically the geodatabase. Why not make the geodatabase open source so that all GIS software could read and write to it? What would be the motive to do this if profit was not a factor?

It's worth mentioning that there were no issues exchanging GIS data between open source GIS software and commercial software. There was undoubtedly more reliability and speed on Linux. The ability to pipe LiDAR data (using *LASTools*) on the fly and into GRASS GIS was incredibly efficient and fast. Command line automation with Bash or Python were really making processes hum right along. It felt so good to be in an environment where the software worked as advertised, on a stable system, using 64 bit hardware and software. It was very similar to the early days of ARC/INFO on UNIX but better. There was an enormous and intelligent community surrounding open source software that made it the best stuff on earth.

8.15   Open Source GIS Exceeds All Expectations

There was enormous success achieved with GIS analysis and modeling using open source software over the years. Using open source GIS for modeling and analysis has proven its value, while working faster and smarter. By now, there was over 310,000 acres of land modeled at one meter resolution for various projects. Using open source software on Linux for GIS analysis and modeling has become a regular practice to meet client requests. This is one important variable that counted as a major success: meeting client requests with all open source software.

Using open source software saved $200,000+ over the years for one user. It also freed up advanced GIS licenses and modules for coworkers. It's not like using this no cost software sacrificed capabilities or performance. All of this was great, but from a user perspective, the best part was that the fun had returned to using GIS. Not only was it fun, but we were back on a UNIX clone, Linux. Not to mention a great community of problem solvers and people willing to help. It was the best of times with so much information learned and applied.

As this section came to an end, it was 2015. So much had been done with open source software that finally the journey to open source software was considered complete. Complete in the respect that commercial GIS software had been replaced with open source GIS software for analysis and modeling. Well, complete? Are journey's ever complete, save the day we die? With creativity and passion, the journey is never over and there is still plenty of creativity and passion left to carry on down the road of this journey. Get ready for some GIS-based Minecraft, lake volume modeling, and a theoretical *geoblockchain* modeled after Bitcoin. Then we put it all together in an open source enterprise GIS.

9. Personal Projects: The Fun Side of GIS

9.1 Work Detour

This chapter takes a detour from the work environment and ventures into some fun activities using open source geospatial software. Hopefully this journey inspires some ideas or new things to check out on your journey. Your journey is the real fun, isn't it? There is a lot of information about GIS that can be learned in an academic setting and there is also a lot to learn from real-world experiences.

They say you should try to find work that is enjoyable and you'll never have to work a day in your life. Guess that depends on the perspective of "work". The work described in this chapter using GIS technology was more than enjoyable, it was really fun. If you find yourself doing GIS activities in your free time, there's a good chance you find GIS fun too.

9.2 Modeling Minecraft Worlds From LiDAR

It might be interesting to know that it is possible to use LiDAR data to build a Minecraft world. A lot more information is available on the web now than when LiDAR and Minecraft were investigated in 2016. Now we have GitHub projects involved with LiDAR and Minecraft worlds, but initially there wasn't a rich pool of information to draw from.

The Backstory

This story started in 2016 when a NJ STEM member searched that popular professional website for someone with LiDAR and Minecraft experience. There was only one resulting profile returned. Minecraft modding was mentioned in the professional profile as well as LiDAR experience. The inquiry was to see if help could be given to convert real-world LiDAR data into Minecraft worlds; worlds that could be played by users. Based on the contents of the book so far, it's probably no surprise that this is entirely possible with open

Figure 9.1: Aerial Photograph & Modeled Minecraft World

source software.

Years before this contact, many hours were spent modding Minecraft software with family. It was a fun way to connect and enjoy a game at the same time. Eventually, this led to (shocking!) running a Minecraft server on Linux. The fun eventually expanded out beyond immediate family and soon some old friends were in the mix crafting and building too. With this server administration, there was modding on the client side and the server side. This was well before Microsoft had purchased Minecraft and wrecked it. Ahh, the good ole days, they really were just that.

At the same time, the professional career was in full swing using LiDAR data to model terrains and perform hydrologic modeling. We've already covered the aspects of LiDAR, so it should be no stranger at this point. The general workflow to create Mincraft worlds from LiDAR data is: obtain LiDAR tiles for the area of interest, filter for ground points, import to GIS, manipulate data to fit Minecraft constraints, paint the world, then export to a height map.

Minecraft Terrain Constraints

The first realization was that Minecraft worlds have a height limit from top to bottom of 256, measured in blocks. One Minecraft block measures one meter long, one meter wide and one meter high. The one meter by one meter horizontal resolution wouldn't be a problem, but the elevation values would need to be adjusted from real numbers to integers.

DEMs are typically in floating point formats because the data is more precise than a foot or a meter. For example, someone may live at a 188.5 foot elevation. We measure elevations in fractional numbers (floating point decimals), but the smallest unit Minecraft works with is one one meter, by one meter, by one meter blocks. The DEM in GIS is built using the same process from LiDAR, but the data does need to be rescaled to fit the Minecraft block size constraints. Once the DEM has been created and prepared, it can be exported to a GeoTIFF image for WorldPainter (described below).

Chosen Model Area: The Beach

The area chosen to be modeled and play Minecraft in was Ship Bottom, New Jersey.[26] Ship Bottom is a beach town on Long Beach Island where many teenage summers were spent surfing. Since beaches are mostly flat, the first issue was encountered. Modeling a flat area with one meter (three foot) blocks means that a lot of small elevation changes, within a meter, will not be represented. Even considering the flat beach and the one meter height interval restriction, the surface modeling results were very useful.

WorldPainter

Now we need to build the Minecraft world. Too bad this is going to require some expensive software. Psyche! This is an open source journey, so forget that. Time to go fishing in the sea of open source software for a solution. Whoa, got one! What is it? Cool, it's a Minecraft WorldPainter.[27]

Worldpainter is free and open source software. The software can import a DEM as a GeoTiff and export the data as a height map. In Worldpainter, the world is 'painted' with the blocks found in Minecraft such as sand, gravel, dirt and grass. This is pretty cool, but it seems like it could be difficult, for some artistically challenged users, to look at an aerial photograph and then paint the features into a Minecraft world. Look over there, memorize the patterns, go over here and paint them into the world. Very tedious and time consuming, but we wouldn't be doing that.

Worldpainter has a neat feature that helps users replicate the real world into a Minecraft world. One very useful feature is the ability to overlay a transparent aerial image. This resource is incredibly beneficial when painting a Minecraft world, because the underlying photography provides all the features that need to be mimicked in the Minecraft world. Using an aerial photograph helped

---

[26]https://www.google.com/maps/place/Ship+Bottom,+NJ+08008
[27]https://www.worldpainter.net/

paint the world faster, but the process was still very time consuming. If you want the detail in Minecraft, you have to paint it in. Painting. Lots of painting. It felt like an infinity of painting. Eventually, the goal to paint all of Ship Bottom was abandoned and the project was left in the state presented here. It was cool and fun, even for someone not artistically inclined.

When exporting, the WorldPainter program makes a Minecraft "save game" that the user later loads in Minecraft. The terminology for elevation data in a Minecraft world is called a height map, what we would call a DEM.

Pressure to Use Commercial Software

Enough work had been done to present a proof-of-concept to the gentleman who initiated the LiDAR Minecraft request. As the workflow and solutions were presented, it was learned that they wanted to use FME for the DEM conversion process. FME is expensive commercial software that does a lot of things and one of them is data conversions. A specific example is converting a GeoTiff DEM to a height map. An FME license was offered for use, but at this point in the journey there wasn't anything interesting about commercial software. Time spent learning about commercial software seemed like time lost because there was always a solution found in open source software. Not to mention the awesome community of users.

At this point in the journey, the belief in open source philosophy was too strong to go forward with a project using commercial software as the solution. Part of the objective was to provide a free solution that any student could use, even outside of the STEM group. Providing a commercial software solution to students didn't seem right. Sure, the commercial company would provide some free or discounted software for the project, but what about other students? Not to mention, that's how this whole commercial trap starts. One license, then two, then three, when really it should have been open source software from the beginning. Ways were parted when this divergence in thinking was discovered, but the experience was cool and created this opportunity to share it.

The Illustrations

The first image shown (Figure 9.2, on the next page) is an aerial photograph of Ship Bottom, NJ. Truly a magical place. Don't tell anyone, but Ship Bottom and Long Beach Island in general have some really great surfing.

Below is the painted, and incomplete, Minecraft world (Figure 9.3, on page 183). The major landforms were created with relatively broad painting strokes. The

Figure 9.2: Aerial Photograph of Beautiful Ship Bottom, Long Beach Island, New Jersey. Courtesy Google Images

grassed dunes, beach, and ocean didn't take very long to create. Painting the features in layers seemed to be the trick, so roads were painted then grassed yards and buildings were placed on top. This process minimized the amount of detail required to paint around objects like buildings. There were paint filling operations and other tools like brushes with different types and sizes. Details like houses and residential features were painted on top last.

WorldPainter makes it easy to not only paint features onto a surface, but also do define biomes that include certain kinds of features. But keep in mind that Minecraft also auto-generates multiple levels of blocks depending on their height/altitude and biome settings. So, while you may "paint" a top layer to look a certain way, below ground may still be quite a mystery until a player ventures below the surface.

Special time was allocated to paint in "The Sand Trap" miniature golf course. If you are a local, can you spot it? It's not really apparent otherwise, but it's the burnt red rectangle on the map. If you go to Ship Bottom, be sure to stop in and play a round. The owner at the time, Larry, gets a special mention. He definitely started the "cool boss" list. This was a great summer job, with quick access to surfing at the beach just a block away. Two good friends worked here, and we were always willing to watch someone's shift for a short while if the waves were pumping. Nighttime at The Sand Trap was just the best of times for working. For the non-locals, a baseball diamond is also painted into the world, can you find it in the image below?

The next image (Figure 9.4, on page 184) is taken from within the Minecraft software. In creative mode, users can "fly", which is why this view is from an elevated perspective overlooking the dunes. There is a path to the beach that goes through and over the dunes. The LiDAR data modeled this feature quite well. They're hard to see, but Creepers are in the dune path.

The next image (Figure 9.5, on page 184) shows buildings in the landscape. The buildings were constructed in the WorldPainter software. There was another neat software feature that elevated features from the surface. There's a fancy word for this but who cares. The buildings can easily be created in WorldPainter with the one meter block size limitations. The surface was a bare earth DEM and it's almost hard to believe we're talking about this in the context of Minecraft. Maybe modern use of LiDAR in Minecraft modeling incorporates the full point cloud, such as buildings heights. Surely there is a way to incorporate this. Maybe you can hop in and create some new tools or think of some new ideas. Let's see!

The next image (Figure 9.6, on page 185) is an oblique view from within the

Figure 9.3: Painted Minecraft World From LiDAR

Figure 9.4: First Person View in Minecraft

Figure 9.5: First Person View - Buildings in Terrain

WorldPainter Software. This change in perspective is another neat feature in WoldPainter. If you couldn't spot the baseball in the previous image, can you see it now?

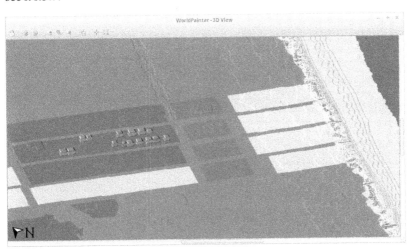

Figure 9.6: Oblique View in WorldPainter

Build on the Idea

Overall, the experience was really fun. The idea to create a Minecreaft world from LiDAR and then play in it is really intriguing. What an excellent educational project, merging GIS with gaming. It was a lot of fun to paint and create the world using different Minecraft blocks. How cool is the WorldPainter software? Pretty amazing open source software.

Taking this one step further, Minecraft can be played in a virtual reality headset. The application of real-world terrains in a Minecraft virtual reality could bring the project to a whole new level. If you like Minecraft, and you have a passion for GIS, you may find this to be a great hobby to keep you busy for a long time. Have fun and please share your creations for others to see and enjoy.

9.3   Integrating Bathymetry with LiDAR Topography

Fun project #2. You know you are having fun with your chosen profession when you are intrigued enough to try something on your own time. While

Minecraft worlds may not have a professional application, the next use case for open source GIS certainly does.

Figure 9.7: Mountain Lake Boat Ramp

Composite Elevation Model

What are we doing today? Going boating! The goal of this project is to create a composite elevation model, consisting of topography and bathymetry. The processes described here illustrate how to gather your own local bathymetric data and integrate it with topographic LiDAR point data. The LiDAR point data will be from a public data source, but the bathymetry will be collected with SONAR equipment.

As a bonus, we'll compare lake model dimensions. Specifically, we'll look at volume differences between a DEM from LiDAR only, and a composite model using LiDAR and SONAR data. It should be no surprise that at this point that we'll be using all open source software. We'll use open source geospatial software to process the LiDAR and SONAR data, as well as perform the volumetric modeling.

LiDAR Data & Water

LiDAR point data is truly amazing. We already covered this point ad nauseum in an earlier chapter. The combination of highly dense and regularly spaced points are the right ingredients to build a high quality DEM. When LiDAR data is flown to capture land elevations, it doesn't capture elevations beneath a water surface. LiDAR points over water often reflect the surface water or provide erroneous results. LiDAR water points are often filtered out when creating a DEM. If the feature is a large lake, the surface water values are interpolated from the lake edge points or replaced with a hydro-flattened water surface.

As a review, a good LiDAR product has standard ASPRS classifications for the point data. This means that only classifications of interest, such as ground, can be extracted from the LiDAR point file and used to create a DEM for areas above water. Three basic LiDAR point classifications are: unknown, ground and water. Sometimes there are other classes, such as roads, rail and high vegetation. There are times that LiDAR elevation values can be used for open water surfaces, but other times they can introduce errors. These situations will need to be reviewed. In this lake modeling application, we're only interested in the LiDAR ground points, so we extract ground classification number two.

Looking at the aerial photography and the LiDAR DEM (shown below), it is interesting to note that these two data sources were collected at opposite times of the year: wet and dry season. LiDAR does not penetrate water (save special equipment) so we need another method. A traditional way to survey lake bottoms is to use a rod or chain and manually measure depths. A modern way is to use SONAR. SONAR uses sound waves to determine distances to objects. We all are familiar with the PINGs associated with submarines and maybe even in computer systems.

Project Challenges

There are two interesting challenges in this project. The first challenge lies in establishing the elevation of the lake when the sonar is collected. This is needed to convert the SONAR depth readings into elevations above sea level so that the SONAR and LiDAR data reference the same starting elevation. Lake gauges can be used to determine water level elevations, however, in this situation the lake guage was not used.

Related to the first challenge, the second challenge lies in reconciling the dry-season topography elevations - the exposed lake bottom - with the wet season SONAR elevations. In other words, we have LiDAR data mapping the land surface and at another point in time the lake is filled up while we float a boat across the same area obtaining SONAR depths. The data needs to be reconciled between the dry land LiDAR and the water filled lake SONAR.

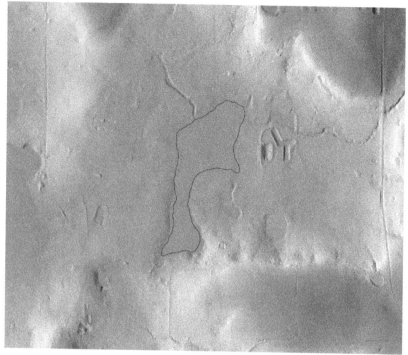

Figure 9.8: LiDAR Only, No Lake Points

Shown in the central-northern part of the LiDAR DEM, there is an area of the lake bottom exposed as land. This means there is LiDAR data inside the illustrated lake edge; an interesting predicament to rectify. For best coverage of ground points, the LiDAR elevation data was likely flown in the dry season when tree leaves were not present.

The aerial photograph above, however, shows the lake water elevations at near full levels (Figure 9.7, on page 186). The imagery from the *National Agriculture Imagery Program (NAIP)* is flown for vegetation inventory.[28] This is typically collected during leaf-on season and corresponds loosely with the wet season. Since the LiDAR captured "land" values inside our lake boundary, we will need to carefully cut and merge the LiDAR and SONAR data to create a composite model; or so it was thought initially. Here's a clue so as to not be so vague, we are looking to map the lake bottom and the LiDAR managed to

Figure 9.9: Aerial Photo with Lake Outline

scan it as land.

SONAR Data Collection

The data collection was without issue and was performed in a Gheenoe with an electric motor.[29] First, the perimeter of the lake was navigated while collecting SONAR points. A distance of about six feet was kept from the shoreline as the edge data points were obtained. Then, the boat made two passes down the longest dimensions of the lake. Only two lines were obtained trying to keep it from turning into work, but also a Gheenoe with three kids during the survey is a challenge also.

SONAR data points are collected with great density, however, it is extremely

[28]https://datagateway.nrcs.usda.gov/

[29]"Gheenoe is the trademark name brand for a long, slender, shallow-draft boat design which is superficially similar to a canoe but is much more stable" (findwords.info)

narrow and linear with the travel of the boat. More sophisticated bathymetry scanners scan with a swath, but we were limited to an exact location straight down below. Remember the problem with topographic contours? It's the same problem. Tightly packed elevation points along a line leaves for a lot of other places without good data points. Having a line with the same elevation to create a DEM isn't nearly as useful as having a bunch of points that are well distributed.

SONAR Data Processing

This was a new thing, processing SONAR data into a GIS layer. We weren't using the regular GIS tools this time, we had to find something new. It was time to go fishing in the open source software world again, what would we catch this time? A line was cast and the next thing we knew, a beautiful piece of software was reeled in called *MB-system*. An open source software for SONAR data. Wowee.

MB-System is an open source SONAR data tool that can process, edit and filter sonar point data. It uses horizontal correction features, such as dead-reckoning, to interpolate points that have erroneous latitude, longitude or heading. MB-System doesn't allow users to create their own points, but only manipulate the points so that they fit good data patterns. This is an intentional design to keep data as close to the collected measurements as possible. It was cool to see that other people believed it was important to keep data as close to the survey values. Similar to the earlier thoughts that commercial GIS software shouldn't have to modify data to work with its software, the data should try to be kept as native as possible.

Once the sonar points have been processed and edited (beyond scope of this discussion), they can be exported as X, Y, and Z points in decimal degrees or an EPSG-friendly projection.[30] In this example, the points were imported into our favorite analytical GIS, GRASS GIS. The Z value captured represents the depth to bottom and still have to be converted to an elevation.

SONAR and LiDAR Integration

Before the SONAR and LiDAR data can be merged, the SONAR point data needs some further editing. First, the SONAR point data needs to be converted from SONAR depth to elevation above sea level. This is determined by using the lake's water level elevation and subtracting the SONAR water depth value.

---

[30]https://epsg.org

How can we know what the lake elevation is when the boat took the SONAR depth readings?

Using the LiDAR data as a basis, a model was used to flood the lake to an elevation that matched the level when the survey was taken. How can we know the lake elevation? A few clues are that the boat's SONAR path needs to be fully contained in the lake model; that is, we have to float on water. Also, it's known that the first pass of data around the edge of the lake was about six feet from the lake edge. A review of lake elevation and SONAR depth values is illustrated below (Figure 9.10).

Figure 9.10: Aerial Photograph - Flooded to 97 ft.

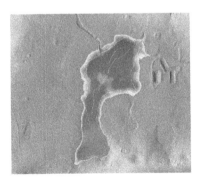

Figure 9.11: Shaded DEM - Flooded to 97 ft.

After the water level elevation is determined, the SONAR depths can be converted to elevations. If we know that a location has a two foot depth and the water elevation is at 97 feet, then we know that the lake bottom is 95 feet at that location. This information can be applied for all of the sonar depths, converting depths to elevations, referencing the *1988 North American Vertical Datum (NAVD88)*. We know the datum reference because the LiDAR data is referenced to NAVD88.

The SONAR equipment collects very dense points in a line; the same problem as the topographic contours. The first thing that happens is binning; many points are resolved to one value per raster cell. So, when converting this data into a continuous raster elevation, the data is smoothed to reduce a local influence of point values. The results won't be awesome, we need a lot more data points for that. Once the SONAR depths are converted to a projection and reference elevation (NAVD88) that matches the topography, the topographic and bathymetric input elevation points can be merged into one file for composite surface interpolation.

An intermediate step performed, that probably shouldn't have been in retrospect, was to buffer the SONAR data 10 feet and remove LiDAR inside this distance. Since the SONAR data was about six feet from the lake edge, the idea was to smooth the transition from SONAR data to LiDAR data edge. Due to what can be sudden bank changes in a lake, the LiDAR data should be used rather than not. Once the points are prepared, a full composite elevation model, both above water and below, can be interpolated. One observation is that the southern part of the lake is deeper than the northern part. Also, the lake bottom scanned by LiDAR is visible in the central-north area of the "Lake Defined by LiDAR" image (Figure 9.13, on the facing page).

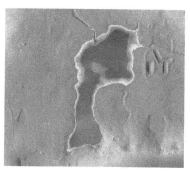

Figure 9.12: Partial Lake Bottom Defined by LiDAR

Using another perspective, an oblique view from the northeast, the dimensions of the lake can be observed. Different Z values can be used to exaggerate features, but this illustration was kept rather modest. Below, are three models illustrated in 3D space: lake interpolated by LiDAR (Figure 9.14, on the next page), composite elevation model (Figure 9.15, on page 194), and finally a model showing the terrain flooded at 97 feet (Figure 9.16, on page 195). While the difference is subtle, the impact of the SONAR data can be seen for the lake bathymetry.

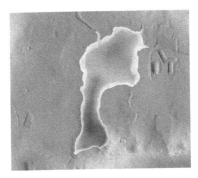

Figure 9.13: Lake Bottom Defined by SONAR

Figure 9.14: Oblique View Lake From LiDAR

Figure 9.15: Lake From Sonar - Note Indentations at Lake Edges

Lake Volume Estimate

The visual illustration of the model differences isn't that impactful, but running the numbers is. Now we have two models that can be used to determine lake volume; one that is from LiDAR with no bathymetry and a composite model that has bathymetry incorporated into the elevation model. If the lake is flooded to 97 feet, we can compare the two lake models.

The lake model that uses LiDAR only, and is in the dry season, has the following metrics: 2.8 feet maximum depth, 42 acres of surface area and 125,807 cubic yards. When the composite lake model is used, the lake volume increases over 100%, with a maximum depth of 9.5 feet, a surface area of 42 acres and a volume of 280,242 cubic yards.

SONAR Wrap-Up

Everything worked as expected: the LiDAR data, SONAR data collection, SONAR data edits, LiDAR and SONAR data integration and volume calculations. The effort proved to show a substantial difference in lake dimensions

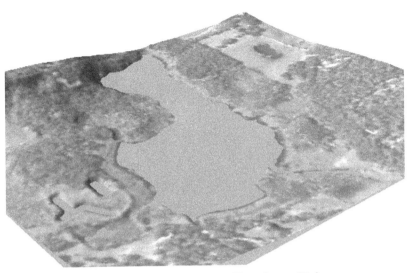

Figure 9.16: Lake From SONAR with Water Elevation at 97 ft.

by gathering bathymetric data. The results are not shocking, but the process is cool for situations that need this information.

There are definitely some ways this process can be enhanced. It may be advantageous to pre-plan the survey transects and use navigation to ensure good coverage of the lake bottom. At a minimum, even if no pre-planning occurs, gather as much data as possible. Using more points to define the lake bottom would build a better model.

In retrospect, it's interesting to consider using the dry part of the lake that was scanned with LiDAR as the actual bottom. Then bring in supplemental SONAR points around the lake bed area scanned by LiDAR. This would have been a far more accurate way to define the lake bottom. That's why science projects are cool, we get to learn and improve the process for next time. The next thing we know, we're getting state agency approval for a new method developed.

Closing Notes on Desktop GIS

These two projects were done for fun and the cool thing is they are within the GIS profession. Are we back to this dead horse of, is it a profession? We might be getting close to the answer, but at the very minimum, it's spatial software. Some would even argue special. How we classify it beyond that is anyone's label. Maybe labels only matter when they are wrong.

This wraps up all the adventure with desktop GIS. Desktop GIS is where all the fun and cool stuff is. It's the analysis, modeling, mapping; all the things that make people go "wow" to GIS technology. No normie outside of GIS hears about abstract database schemas or apache web servers and gets all wide-eyed with excitement. Desktop GIS can do a lot of neat things and we only scratched the surface.

As we put desktop GIS behind us, we now enter a different area of GIS technology: GIS database servers. GIS database servers are the GIS technology behind the scenes that have the ability to power an entire organization. In this example, we'll look at an open source GIS data library implementation feeding data to that oh-so-dreaded commercial desktop software. The integration with QGIS is not worth a chapter; it just works as advertised.

10. Enterprise GIS Database Library

10.1 Venturing Outside Desktop GIS

The previous chapters illustrated some pretty powerful applications of GIS desktop software and a heavy dose of modeling and analysis. With GIS analysis and modeling behind us, we venture into systems, specifically a database that is enabled to store GIS features. After all this time in the open source journey, it was time to stop avoiding PostgreSQL and PostGIS, and time to start putting them to use.

10.2 Parallel Journey With Linux

Throughout this open source GIS journey, there was a parallel journey with Linux. In the same way open source geospatial has its own domain of software and data, so does Linux. Using Linux as a desktop workstation is quite easy today, but using Linux as a database server or web server requires a little more knowledge. Compiling open source desktop software requires knowledge about components of Linux, and more specifically, other software libraries. In all of this, the operating system is the glue that binds the geospatial software, services and data together.

The focus of this journey hasn't been on Linux, but in the background it is the underlying sandbox providing great play time. All the other software, like geospatial, were just toys in the sandbox. They are toys in the sense that they are really fun to use and they do interesting things that occupy one's time.

It was now 2015 and three industry standard Linux certifications were obtained to further knowledge about the Linux operating system. At the time, one exam would provide certifications from three institutions: LPIC-1, Suse and CompTIA Linux+. There were several reasons the certification was desired. The primary reason was to fill in knowledge gaps that were required to administer a system providing GIS services. Another reason was to supply credentials for cloud or other services, and yet another reason was to show that Linux wasn't just a hobby anymore. Linux had reached a level of offering professional services.[31]

After passing the Linux exam, which was a lot of information but not exceptionally difficult, there was a passion to take open source GIS to the next level. What could we do next? How could we do something beyond desktop analysis and modeling that would benefit the organization? It was clear that the organization needed a centralized data system. A new goal was set: create an open system to share and distribute GIS data layers to all users in the organization.

10.3   Enterprise Problems to Solve

There were two enterprise-level problems to solve. One of these problems slowly becomes an infestation as an organization grows. It lurks and consumes space until IT staff start screaming and yelling. This entity is especially menacing to users who don't know the secret filing system. This glob of trouble is none other than the piles and piles of GIS data that gather up over time with GIS usage. Isolated piles of data with unique user catalog systems end up being stored at different office locations, with different years and versions of data never to be touched again. Even if organized or cataloged, these piles of GIS files can't effectively be shared across a wide area (office to office) network.

The cure to this menacing infestation can be found in an enterprise GIS data library; a centralized database source with read-only, widespread data coverage and system redundancy for ultimate uptime availability. This GIS data library would be a single place GIS users could go to obtain clean, ready-to-use GIS data.

The second problem to remove was access barriers to GIS data. GIS data was heavily confined to commercial software lock and key. For example, suppose someone wanted to know if a wetland was on their project site. They had to go through a special GIS user, with special software, to display a special file format. Why can't they bring up a web page and retrieve this information themselves? Or add a file to their Google Earth session? These users just wanted to use GIS data like a library book; look, review, and return. If the data could be in an open and self-serving format, it would provide workflow momentum for both the person requesting information and the occupied GIS worker who would typically be interrupted to fetch the data.

The third problem was interoperability with commercial GIS software. The

---

[31]https://blog.morphisec.com/linux-cyber-security-gartners-cwpp-insights

open source GIS data library had one unfortunate yet essential requirement: it had to work with ArcMap. The commercial GIS company advertised that PostgreSQL and PostGIS tables could be read by ArcMap. For the GIS data library, only read capabilities were needed to clip data from the library for an area of interest. In earlier versions, users had to move the SDE .dll files into an ArcMap directory, but later the SDE files were included with ArcMap, giving it some flexibility in data formats. All indicators were suggesting that this idea could be put into practice.

For the ArcMap users in the organization, there were at least a half dozen layers used in everyday maps. Common layers such as parcels, land cover, national wetlands, streams, water bodies, parcels, floodplains, soils, and topography. Instead of each user in an office downloading and storing all of these layers and many others, the layers would be stored in a single source database container. This single source library would simplify and unify the disjointed nature of GIS data. Per office, some of the data differed by date, versions, source and table fields. With this library system, users could throw an ArcMap polygon at the library and rapidly get GIS layers in return.

10.4 Going for It

The excitement could no longer be contained and the high from passing the Linux certification exam had not worn off yet. Eventually this system would evolve into a dual system with replication, but for now, the PostgreSQL and PostGIS software would be installed on the Linux workstation. We just needed to get it going and like most projects, the rest would work out. Manuals were read, installation configurations were tried, failures occurred, retries were attempted and then success. The PostgreSQL database was installed and configured with PostGIS.

QGIS had no problem browsing the PostgreSQL database and PostGIS tables, in fact that is what QGIS was invented for.[32] The GIS layers were organized by schemas in the database and QGIS could browse this data structure with ease, using a standard tree dropdown browser. The performance was outstanding. Maybe not entirely surprising since PostGIS was one of the original data sources supported by QGIS in 2003.

Now let's make it work with that dreaded commercial GIS software we left many years ago. Can you guess what happened next? Probably. There has always been great confidence in the audience. Well, surprise-surprise, as soon as ArcMap was brought back into the journey, the utopian dream of open source feeding all technology was shattered and jubilation quickly turned to frustra-

---

[32]https://spatialgalaxy.net/2022/02/18/qgis-then-and-now/

tion. It can't be escaped. This plague of non-cooperative commercial software was a constant party wrecker.

Sure, the commercial software advertised that it could read open source database tables, and it could, but in practice the capabilities were severely encumbered for an efficient workflow. We couldn't add a half dozen extra clicks and a humongous loading delay into users' workflows and expect smiles in return. ArcMap was slow and sluggish enough as it was and didn't need any additional processes slowing it down further. It felt like the capabilities were advertised just to meet some open data requirement, rather than actual integration that provided user benefit. Integrating ArcMap desktop with an open source PostgreSQL/PostGIS database became such a misadventure, that it was eventually documented in a paper in 2015, then published on that professional networking site.

10.5 Saving Time and Money With GIS Software Investments

GIS Data is an Asset

There are two common problems with geospatial data: locked proprietary data formats and data accessibility. This chapter provides a solution to leverage *Free and Open Source Software (FOSS)* within an already existing ArcMap desktop environment to create open, accessible and searchable GIS data. Open geospatial data creates flexibility for an organization to choose software that best matches its needs, rather than being forced to buy expensive software to read a specific vendor specific file. Organizations without the knowledge of FOSS get married to commercial software and suffer financially through the long and expensive relationship. As more pieces of commercial GIS infrastructure are purchased, more money is spent every year maintaining it. The cycle is vicious and never ending.

GIS data is an incredible asset to an organization. The asset's value is maximized through open access and distribution throughout an organization. The asset's value is reduced when it is scattered around a network and only accessible by a few operators with expensive, specialized software. The value of the asset is fully recognized and leveraged through serving the needs of an organization.

When geospatial data is kept in a proprietary or locked data format, organizations are required to purchase corresponding commercial software to read and write to the proprietary format. This commercial software is often very expensive and often forces organizations to fall into a vendor lock-in trap. Vendor lock-in makes a customer overly dependent and unable to use another vendor's software without incurring significant costs to change vendors. Vendor lock-in is often accompanied by unnecessary capital expenses. Locked data formats reduce the value of the geospatial data as it becomes inaccessible by other software vendors. As a result, organizations achieve vendor lock-in and become unable to take advantage of the far more attractive open source GIS alternatives.

Believe it or not, accessibility is a common problem within organizations regarding their geospatial data. Imagine having 2,000 Blu-ray movies in 10 boxes. Then imagine wanting to watch a movie. Without searchability, cataloging and automated disc retrieval, all efficiency is lost trying to locate the movie. Now imagine all the movies in a digital, cataloged library that can easily be searched and viewed by pressing the play button. The process to get data goes from 10 minutes to 10 seconds. This movie analogy directly relates to the organizational benefits of organized and accessible GIS data.

Access to spatial data has immense strategic implications with staff at all levels of organizational involvement. Access to geospatial data broadens the pool of customer solutions as more members of the organization can apply their own creativity. Open data access means more members are included in the problem solving process as organizational brainpower is synergized. Organizations become healthier, more efficient, transparent and smarter as access to GIS data is within a few clicks of a button. As a result, organizations are able to serve their customers better and add to the bottom line.

Problems With Centralized Proprietary Data

Proprietary GIS data formats are public enemy number one. First, they discourage interoperability and data sharing. Second, they require the purchase of expensive software to read and write special commercial file formats. Last, but not least, proprietary data formats create bottlenecks as only a few GIS operators in the organization can access and provide data to the organization.

Spatial data living in a locked jail, with only a few individuals possessing the keys, prevents organizations from maximizing the value of their GIS data asset. The abstract objective to free spatial data can be manifested in a FOSS GIS data library. When file-based GIS data is confined to a network share for a few GIS operators, accessibility is lost to others without the key. Knowledge and creativity become isolated behind a wall of expensive commercial software. Other departments that use GIS software may circumvent the wall and collect their own GIS data. This can create fragmented and non-standardized products from within the same organization.

When GIS data is collected within an organization, it is typically stored on a filesystem residing on a shared local network drive. These files can assume several common commercial formats, such as shapefiles or File Geodatabases. Data can be collected from federal, state, regional and local agencies. Since GIS data can be very dense—lots of X,Y vertices—it can be very heavy data to distribute over a WAN. This creates islands of GIS data, isolated to offices, within an organization.

This filesystem-based approach to storing commercial GIS layers creates multi-user access constraints, isolated data collection and knowledge. When remote office locations collect duplicate data, organizations spend more money on storage devices for unconsolidated data. Duplicate data with different versions will create data inconsistency within an organization. Data isolation is a serious problem, as GIS data users can't know what data is available within the organization or which network has what data.

GIS files on a network filesystem cannot benefit from searchability. While names of files can be searched using filesystem tools, keywords within the closed proprietary GIS file cannot be searched. The entire system becomes dependent upon GIS operators' knowledge of what is on the system. An interface or spreadsheet can catalog either file based or table based data, but databases offer an additional level of searchability.

Solution: Free the Data

Geospatial data, in the simplest form, is X,Y coordinate data that models point, line, and polygon geometry. The 'geo' in geospatial indicates that the data uses a *spatial reference system (SRS)* that positions geodata to locations on Earth. A GIS data library, as described here, is a software container with high-tech database properties, that stores and distributes geospatial vector data: points, lines and polygons. The GIS data library is also database technology that is file system and operating system agnostic; the library serves spatial data from a network service.

GIS layers in the library cover very large areas of statewide seamless GIS data. The data library groups large GIS layers by themes so that users can quickly sort through a catalog to find the data they need. For example, the "inland waters" theme may contain statewide layers of floodplains and water bodies. Thematic or topical grouping, with statewide extents, allows for rapid data browsing and increases efficiency for the GIS processes.

During the life of an organization, it becomes apparent that there is value in aggregating spatial data into a single storage container. The container is a system that provides order to the otherwise chaotic data. The system, a *Relational*

Database Management System (RDBMS), is far better at managing, indexing, storing, searching and distributing data than humans. When these attributes are applied to large volumes of geospatial data, it creates intangible value for the organization. Maintaining geospatial data from a single source container is easier than maintaining GIS files on different networks and the same is true for backups. The GIS library adds value to the GIS data as it is unconstrained by local networks, providing GIS data to all network office locations and users.

Once a GIS data library exists, accessibility can be extended to services other than GIS desktop usage. One example is a web mapping service presenting GIS layers to non-GIS users through a simplified interface. Web mapping services are great at illustrating where layers exist in an area of interest. They're also great at querying features and displaying database information to users. These value add-ons provide a competitive knowledge boost to organizational members compared to organizations without this technology. This enterprise-enabling GIS technology raises awareness and ensures single source data usage. Knowledge becomes a global resource with rapid access to have efficient operations.

Another clear benefit is that a GIS data library helps utilize existing staff by providing access to GIS data for basic tasks. Open data reduces dependency upon asking a few members with special software to retrieve some information. Spatial data is not so special that it needs expensive commercial software to access it; remember it's just X,Y data with a table. As members gain the ability to browse GIS data, it helps organizations manage staffing needs for lower-level data browsing and printing. It also allows GIS professionals to expand their expertise by offloading basic data inquiries to members of the organization.

Open Source Benefits for Organizations

Open source software defies the saying "you get what you pay for". Leveraging the power of FOSS creates a unique opportunity for organizations using commercial desktop software to immediately add compatibility and value at no cost. Organizations can keep their current ArcMap software for their specific needs and still use high-performance spatial database technology for their core library data holdings. The FOSS technology can effectively distribute spatial data layers to desktop software, web mapping services, field units, and replicate data to the cloud. Building a FOSS GIS data library can immediately enable GIS data access across an organization's entire network and has the scalability to move to the cloud.

Over the years, the interoperability gap between commercial GIS software and Free and Open Source Software has been closing rapidly. Enormous cost savings with open source software include:

- Open data format and distribution
- Data consolidation
- Open Geospatial Consortium standards
- Reports through tabular summaries
- Vector and raster spatial analysis
- Efficient storage and retrieval
- Scalable architecture to cloud
- Pluggable into other services such as Web Mapping Service (WMS)

These are all valid reasons to see how many pieces of FOSS can be used to minimize capital expenses for an organization and use the best available technology. One can argue that FOSS is better software overall because it is driven by real-world needs rather than a profit motive as exists with commercial software. It is extremely easy to get caught up in the purchasing of commercial software, only to realize that holding such a high volume of licenses is a huge expense to maintain. It is easy for a firm of less than 200 employees to acquire so many licenses, that their yearly maintenance costs could easily be tens of thousands of dollars just for desktop software.

Value Proposition

A FOSS GIS data library provides many opportunities to add value to an organization. Many of the benefits are directly attributed to the organization. Other benefits relate to technology synergy, while other benefits go directly to users of the geospatial data. Some benefits have overlap, but they are grouped below where the primary benefit lies. All of these benefits contribute to a healthier, more efficient company, able to serve itself and customers far better than a company without a GIS data library.

Organizational Benefits

- No software costs; instant value added
- Eliminates bottleneck of only a few select members being able to access GIS data
- Allows for rapid response to customer needs
- Streamlined workflow efficiency

- Standardized symbology for consistent cartography; organizational branding
- Data and GIS knowledge is a centralized resource for organizational members
- Competitive edge
- Empower members of organization with free access to knowledge
- Limit the need for dedicated GIS staff to perform basic low level GIS data browsing
- Realize different usage perspectives (e.g., marketing, scientist, engineer)
- Efficient storage of spatial data saves organization money for hardware storage
- Organization boasts in marketplace of spatial database technology; technology edge
- Cloud deployment makes possible to sell library usage to customers

GIS Team Gains

- GIS data is searchable by keywords and metadata documentation
- Provides an organized system to browse data, grouped by themes
- ArcMap cartography is streamlined through drag and drop stored symbology
- Scale dependent rendering, pre-stored labels and subqueries
- Perform geospatial overlays and analysis with drag and drop capabilities into ArcToolbox
- GIS Team benefits from same consistent and reliable data source container
- Increase in process efficiency and reduction of errors through use of data standards
- Enable organizational members to work with GIS data through lightweight, simplified web interface

Technology Synergy

- Library provides open format for interoperability with commercial software such as ArcMap
- Runs as a service on a network, not single files on a filesystem

- Direct integration with other geospatial software for various geospatial services

- Provides enterprise GIS characteristics through connectivity with Web Mapping Services (WMS)

- Scaleable through server replications and cloud deployment

- Spatial data is stored and served with ultra-efficiency, capitalizing on RDBMS strengths

- Flexibility in replicating and migrating to cloud

- Multi-user access

Financial Analysis

A cost comparison was made between an open source software solution and a commercial software solution (Figure 10.1). Estimate breakdown included: Microsoft SQL Server ($9K + $2500 annual); Windows OS - $880; ArcGIS for Server (multi-user access, web serving - $10K + $3K yearly. With note that ArcGIS for Server does more than what was just mentioned). About $20K initial investment with $5,500 annual maintenance contract. Estimated five year cost of ownership is $42K.

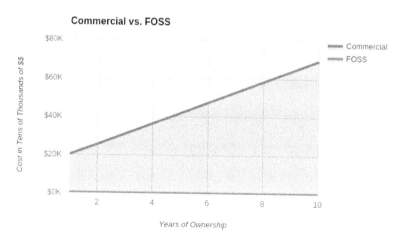

Figure 10.1: Spatial Database Software Costs Over 10 Year Period

Eventually, these services could migrate to the cloud. Estimates for this were not calculated, but renting cloud server space is not cheap. Consideration for offloading this effort to a cloud service is interesting. WFS is a common way to distribute feature geometry through a web serving format. With more remote workers being commonplace, a data distribution format should take into consideration the library's extendability to users outside the LAN or WAN.

Time Value: Efficiency & Customer Savings

The concept of time value can bring significant benefits to organizations and customers by making things more efficient and saving money. It means that by understanding the value of time, organizations can improve how they do things and save resources, which in turn can lead to cost savings for customers.

Consider, a model where there are 10 GIS software users, six producing high-volume cartography and four using GIS data for analysis. GIS users in high cartography production have the greatest efficiency gains. Other users consider GIS data for scientific analysis or engineering applications. While these users are not producing high-volume maps for reports, they are actively working with data for problem solving applications.

For layers, the data could be added from a FOSS GIS data library, symbolized and labeled in as little as five seconds navigating ArcCatalog in ArcMap. Locating data on a filesystem, ensuring the data is current, then symbolizing and labeling could take about four minutes.

A high volume cartographer could generate 10 maps per day using two GIS library layers per map, equating to 20 layers per day or 100 layers per week. Our scenario has six cartographers at $100/hr. Adding about 100 layers to ArcMap a week with a 4 minute time penalty yields a time penalty of 6.5 hours times $100 per hour or $650 a week per cartographer. There are six users, at $3,900, per week.

An analyst, engineer or scientist may use about 20 layers per week at $150/hr, depending on the situation. About 1.5 hours a week saved, times 4 users, equates to 1.5 hours times $150/hr or $900 per week. The four analysts save $46,800 per year and the six cartographers' time value over a year is about $203,000 for a combined time value of $249,800.

Nearly a quarter million dollars in time value can be attributed to implementing a FOSS GIS data library for a small collection of 10 ESRI ArcMap users. Time value grows exponentially as the volume of spatial data grows and more users are provided easy access to the data.

Pathway to Interoperability

A *Relational Database Management System (RDBMS)* is the technology used for storing, searching and manipulating data with multi-user access. Spatial extension software models geometric objects in a database. These extensions are referred to as spatial databases and work within an RDBMS. They can perform spatial analysis using SQL and do all kinds of interesting things that commercial GIS users believe only to be reserved for expensive commercial desktop software.

The FOSS GIS library consists of a software stack. A software stack is a collection of software that works together to provide some service. Open source software and geospatial data standards have made it possible to serve and store geospatial data on a completely open source software stack. This service for geospatial data can be utilized by a variety of client software, including ArcMap. At the foundation, this open source software stack rides on a fast, flexible, robust, and secure operating system that has no monetary costs. Expansion and scalability to the cloud can be done without additional OS costs. It is no secret that this foundation is set upon Linux.

Within Linux software repositories, there is access to many packages that provide geospatial capabilities. One such example is geometric overlays with GEOS (Geometry Engine Open Source). PROJ.4 provides a cartographic projection library to define GIS layers with well-known and widely used descriptions. If the client software can project data on the fly, the data is even more seamless in the workflow. Linux and accompanying geospatial packages provide a well integrated stack of geospatial software that easily competes at the commercial level.

PostgreSQL is the world's most advanced open source database. PostGIS is a spatial extension for PostgreSQL; it enables functionality for storage and manipulation of spatial objects such as points, lines and polygons. Client software, including Arcmap through *Spatial Database Extension (SDE)* libraries, capable of reading the PostGIS *GEOMETRY* type can access and visualize the data from the database.

The PostgreSQL and PostGIS software combination provides far more than the ability to store and serve geospatial data. It is entirely possible to perform spatial operations and overlay from within the database using SQL. The SQL can be submitted from a webpage or GIS desktop application that allows back-end processing. Since ArcMap is distributed with SDE libraries that can connect to a PostgreSQL database and utilize PostGIS geometry, the concept of the FOSS GIS Data Library becomes reality.

Proof of Concept Defeats

The initial objective was to get ArcMap 10.2.2 to browse a PostgreSQL database by schema and then to load GIS layers. The schemas were used to create thematic groups of data. For example, under a 'boundaries' theme there could be GIS layers for: cities, counties, and states. Using the `shp2pgsql` command and gui tool, shapefiles were loaded into the database (Figure 10.2).

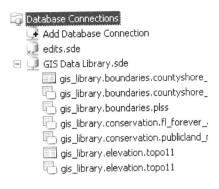

Figure 10.2

The first problem encountered was that ArcMap had no schema browsing capabilities. The browser just displayed each feature once and the table once as a fully qualified object—database.schema.table. Three schemas are shown on the right: boundaries, conservation and elevation. This is really peculiar; they knew that tree-style browsing is expected behavior, it's in their geodatabase. Yet they still offer a sloppy presentation of PostGIS tables while advertising compatibility.

Figure 10.3

The next problem encountered was the listing of a table and a feature for each PostGIS geometry type (Figure 10.3). This is stated to be fixed in ArcMap 10.4. Commercial tech support stated that the software was not able to read the constraint type for a geometry (e.g., `geometry(Point)`), and as a result there were two listings: a table and feature. This double listing of the layers made it very confusing for users to browse the data in a quick manner; almost a deal breaker. The library concept was beginning to look useless in practice with layers listed twice and no way to browse by thematic schema.

When a spatial database feature is added to ArcMap, it is added as a *Query Fea-*

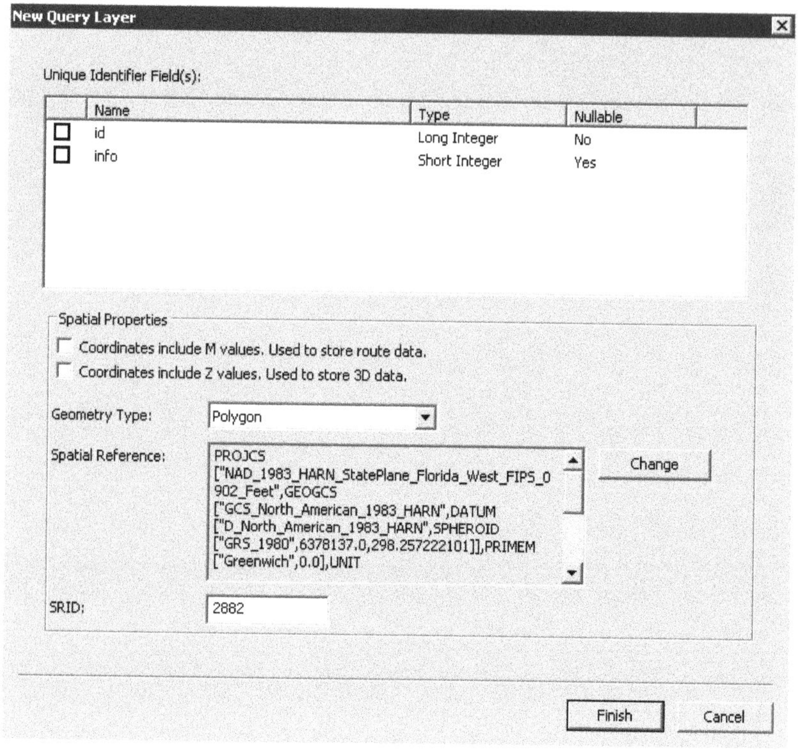

Figure 10.4

ture Class (Figure 10.4). There is a pause in the workflow as ArcMap prompts the user to identify the primary key for the PostGIS layer. Primary keys are important to create unique identifiers for spatial features. Notice the spatial reference is read from the database library. A well-known projection is a huge benefit to GIS software, so that it can project data on the fly and into other systems.

Another pause in the workflow came from the calculation of the spatial extent for the Query Feature Class (Figure 10.5, on the facing page). This took a while, 5-15 minutes for large state-wide layers and was ultimately the deal breaker. For small and simple geometries, the extent was quickly calculated. Since the intent of the library is to have statewide GIS layers with large extents, calculation of the spatial extent can take a very long time. There is the option to manually input the extent (not feasible for production work) or use the *Spatial Reference Extent*. The latter is not optimal, but does allow the user to proceed quickly to the next step of the data being loaded.

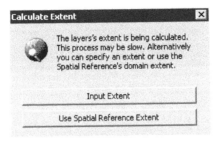

Figure 10.5

There were now several aspects of the commercial and FOSS integration that caused concern for the usability of the library. The layers need to be drag-and-drop into ArcMap, with at least the same ease as a shapefile or Geo-database Feature Class. The inability to browse by schema, duplicate layer listings, more clicks for users to add data and long pauses to calculate spatial extents would not be welcomed by users. Even with the data aggregation value added, this workflow would slow the user down and overall not be a benefit. Bummer.

After several discussions with experienced commercial technical support analysts, a few things were clear. There would be no browsing by schema; a feature request was logged with no hope of ever being fulfilled. The commercial company proposed a preposterous workaround solution that involved having a database connection for each schema to browse by theme groups. This would involve creating a role to own the schema and then map that user as a database connection. This creates dozens of database connections to mimic schema browsing and would require backend database revisions to be done by the spatial database administrator.

Note that because the PostGIS extension for ArcMap was closed source, it meant that community members could not improve its design, performance, utility, etc. Another reason to build on an open and extensible architecture that many open source software projects rely on.

The double reading of layers was a commercial software defect that was stated to be fixed in 10.4. The issue could be fixed by going into each table and altering the geometry types to be unconstrained (e.g. geometry, vs. geometry(Point, 3087)). More back-end database edits to accommodate defective commercial software. Even with those fixes, there was no automated way to choose the primary key and there was no way to have ArcMap read the spatial extent from PostGIS rather than compute it. There would be added key clicks in the process, which defeats the objective for an efficient work flow. This idea seemed dead and unusable.

Then that magic moment happened. On a call with tier two technical support, the commercial support analyst asked a profound question, "How about using layers to build the library?" Making a directory of .lyr files in an organized file-based system may not sound like a practical solution because it calls for an unorthodox implementation. However, the analysis proves that the benefits gained to an organization are innumerable when using the prescribed implementation for commercial software integration.

Proof of Concept Success

Layer (.lyr) files are binary reference files that only certain commercial software can read and make use of. Not to be confused with common "GIS layer" terminology, such as land cover, soils etc. Layer files are used to save parameters about a GIS layer so that when loaded into ArcMap the layers have pre-defined parameters set. Layers are a separate file that reference a GIS data file. Layer files also create a standardized way to symbolize GIS data. Layer files can be applied to GIS files such as: geopackage, shapefile, geodatabase, ARC/INFO coverage or Query Feature Class.

All kinds of useful parameter settings can be stored in a layer file. An added bonus is that they are portable. In the database context, they are system path independent because GIS tables are referenced by IP and port, not a shared filesystem. Someone can make a layer file pointing to a GIS library layer in one office and then send that layer file to another user in a different office location. This portability is possible because the spatial database—a GIS data library—is running as a service on a Linux server. It's important to always recall that layers are a reference file to another datasource; this datasource is a PostGIS table in a spatial database, being provided as a network service.

Layer files became the perfect solution to merge commercial desktop GIS software with an open GIS data library. When GIS layers are cataloged by themes or topics, it creates an easy and efficient way for users to find data they are looking for. Even if the layer file did not solve the problems described for usability, they make for a strong case to build the layer library due to the benefits below. Note, however, these layer characteristics are specific to one commercial GIS file type (.lyr) and do not work with other commercial GIS software or open GIS desktop software. Interoperability of the file format is lost with other systems. This is fine because QGIS doesn't need these files and interoperability was lost the moment vendor-locked commercial software was chosen to be the GIS desktop software. This interface, through layer files, allows ArcMap to co-exist with a Linux FOSS GIS data library.

Immediate Benefits of Layer Files for ESRI Software Users:

- Fast drag and drop functionality; huge productivity gain
- Easy to share complete library to all commercial desktop users
- Store and auto-symbolize data with styles
- Spatial extents are quickly read, not calculated
- No prompt for primary key
- Performance enhancing, scale dependent rendering presets
- "definition queries" in ArcMap are saved, allowing for subsetting of data from complete master layer (e.g., `landuse='wetland'`)
- Discards defective double feature/table read from ArcMap
- Labels can be preset; including font, halo, shading, offsets, location, color
- Many items in the Layer Properties of ArcMap can be stored in a layer file
- `.lyr` files are small in file size, easy to transport to users
- Entire library of layer files in directories can be zipped into one file and quickly distributed to other users
- Brands GIS data to organizational standards through common symbology

Loading Spatial Data Into PostGIS

The process to load data into the PostGIS data library is relatively simple. The database tools `shp2pgsql` and `shp2pgsql-gui` load shapefiles into specific schemas in a database. Attention should be focused upon the importance of ensuring correct geometry and that useful tables are loaded into the library. The library is the final resting place for the data and it needs to be topologically correct. At a minimum, layers that are loaded into the library need to be checked for overlapping features or incorrect gaps.

Other aspects of the data to be loaded are feature type (point, line, and polygon) and spatial reference identification. Metadata, usually in XML format, can be put into the database to create high searchability. Other columns can be added to the tables to create keyword searches. Many GIS layers can be obscure because they can have a complex attribute table. Metadata helps users get exactly what is needed.

.lyr Creation

After the data is imported into the PostgreSQL database as PostGIS spatial objects, permitted users can connect to the database and select a PostGIS feature to load into ArcMap (Figure 10.6). Upon dragging the PostGIS tables, users will be prompted for the spatial extent. A command line tool `ogrinfo` can be used to get the extent of the PostGIS table. This extent can be assigned to the table when it is added into ArcMap. The extent will not be asked for again, since it is saved in the .lyr file.

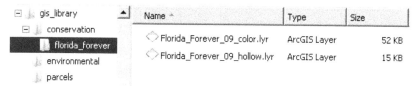

Figure 10.6

Once the layer is loaded into ArcMap, the user can open the *Properties* dialog box to set symbology and other properties. Then *Save to Layer...* from the context menu will save a .lyr file to a specified place on the filesystem. These layers get filed into appropriate directories. The directories are structured by state and then by ISO19115 topic categories. Different symbology (hollow, solid, transparent) can be created for different uses, increasing productivity as most common varieties of data illustrated are already captured in the layer file symbology. These layers can also be used to create symbology standards for data after it is clipped from the library and presented on maps.

Layer Usage

Once a layer is loaded into ArcMap, it can have its properties changed to suit other needs; the symbology and attributes from the layer file are not permanent. After creating these .lyr files, what emerges is a directory of organized layer files, which becomes the ArcMap interface to the FOSS GIS Data Library. The entire root `gis_library` folder can be zipped and distributed to anyone in the organization since they reference PostGIS tables in an RDBMS. The layer files have been shown to be very small and easily transportable. The library layer files can be unzipped for other commercial software users to immediately start using the library functionality. This has shown to improve productivity, improve the user interface experience and decrease time to market when using a distributed GIS team over a WAN. For example, this is especially useful when a small distributed team is quickly formed to respond to a *Request For Proposal (RFP)*.

Concluding Thoughts on GIS Data Libary Implementation

Geospatial data is a heavily relied upon resource and a valuable asset to an organization. The value of this GIS data asset is determined by how the data is organized, managed, indexed and made openly accessible to empower members of an organization. Integrating a FOSS GIS data library provides open and flexible GIS data, saving an organization valuable time and money. Using .lyr files as the interface to a FOSS GIS data library provides a clean and elegant solution to many problems faced by commercial vendor lock-in. These layer files create a seamless integration of the FOSS GIS data library with ArcMap allowing users to quickly and easily access GIS data layers at a statewide level.

With technology, it is important to always be mindful of the future. Maintaining GIS data in an open format provides the best flexibility for an organization to adapt to changing needs and technology. Organizations that buy into vendor lock-in, end up surrendering profits every year in maintenance fees. Organizations that do not succumb to vendor lock-in are able to utilize optimal software that fits their internal needs or customer needs without carrying expensive yearly costs. They're able to change and adapt as technology does in the marketplace. Maintaining GIS data in a centralized, multi-user accessible and interoperable spatial database gives organizations an enormous competitive edge compared to organizations manually filing data on a network filesystem, in closed proprietary data formats.

Going back to the Blu-ray movie analogy, managing GIS data in an RDBMS library gives an organization the ability to browse and press *play* on their favorite data instantly. Meanwhile, competitors without this technology are rooting through thousands of layers, in different boxes, with no index to speed up the searching. One of these organizations will suffer great monetary losses and have severely impaired efficiency. The other organization will benefit from high quality, free software that promotes a lifetime of happy and productive users, who are empowered with knowledge through an open and accessible FOSS GIS data library.

10.6 Fast Forward Five Years

Fast forward five years from when this GIS Data Library was created. The system had proven to be exceptional in every way: robust, fast, and always available. It had, in fact, promoted happy and productive users. There were no monetary expenses whatsoever, as the database servers were recycled from previous GIS desktop use. There was no down time for the service. Well, okay, there's a technical definition of down time and it was down in the sense it was powered off. However, in all fairness the *Uninterruptible Power Supply (UPS)* was on borrowed time. The power outage(s) had occurred during non-business hours, so no users were impacted.

Remarkably, it's worth noting that the two servers used for deployment were several years old and donated by IT staff for the project. It was not special order high performance hardware. From a user perspective, the system was available everytime it was needed; happily serving data to anyone who wanted to consume it. When there were issues, it required restarting ArcMap, not the database. The data library was flawless in the sense that the software and Linux systems did not hang up or crash; they provided constant service to a wide range of users. As far as impact to the organization for downtime, there was none in five years. Impressive.

10.7 Redundancy Through Replication

As the GIS Data Library grew in both the amount of data it was holding and the number of users, it became clear that taking precautionary measures against failure would be very wise. As users came to rely on this system on a daily basis, downtime became a concern. Precautionary measures were needed to ensure high-availability. Common failsafe measures include mirrored drives or dual power supplies in one system.

After exploring options and considering that both systems were already dated, it became clear that replication was the best option. Replication is having an exact copy of the database system running in sync with the main system. If any component failed, such as power supply, hard drive or cpu fan, there would be another copy of the system running. Also, if the primary database died, the stand-by or secondary system would assume the primary role. This automatic escalation of the secondary system to primary status is known as failover.

With the current design, the penalty for using a replicated system was having to create two sets of .lyr files. In practice, it wasn't that much more effort because once the spatial extent of the PostGIS table is retrieved, the extent values can be applied to both layer files. Making two sets of layer files was cumbersome, but the redundancy was a great way to reduce emergency response anxiety. Ah, the price we'll pay for a good night's sleep.

There were plans to implement pgpool-II, to balance the load and use one IP.[33] In theory, this would only require one .lyr file since the software would choose which database to retrieve the data from.

---

[33]https://www.pgpool.net/

Ultra High Performance System

Performance was outstanding. It was faster than accessing data on a LAN server. The system also made access from remote offices possible, where otherwise impossible to do with file-based GIS layers. The true performance of this system wasn't appreciated until working through the problems of *shape integrity error* (more on this below) or some related commercial software issue. While on a shared desktop with the Tier II enterprise expert, he was astonished at the speed in which the statewide layers rendered. The spatial extent filtering on the server side was remarkable and he commented "I have seen a lot of enterprise setups, but never one this fast." He couldn't believe it was running on old hardware, but after all, it was Linux. At this point it was obvious we had something very special.

A performance test was done with a state-wide GIS layer, one as a PostGIS table and the other as a shapefile. When rendering the data at a local scale in ArcMap, there was a 104X performance difference. The PostGIS layer rendered in 1 second vs 104 seconds for the shapefile. As far as data extraction from the library, clipping the data to a project extent had a 7X difference in speed; favoring the PostGIS tables. For organizations needing access to statewide GIS data resources, it was clear that this system was best in class.

Data Pre-Processing For ArcMap

Issues that were encountered are directly related to the ArcMap upgrade at 10.5. After upgrading ArcGIS desktop clients, a common error started to emerge for some layers: *Shape Integrity Error*. Sometimes partial data would render in the display or partial data would be provided in a clip output. The commercial GIS company said they removed some data checks in the code and recommended the "repair geometry" tool. Oddly, in ArcGIS there was only an option to repair geometry using their software validation, but in ArcGIS Pro, there was another option, "OGC" (Open Geospatial Consortium) validation. It was OGC validation that needed to be applied to the data, if data was to be served from PostgreSQL to ArcMap. Sure is kooky, but OGC validation fixes it.

Most of the errors were related to duplicate vertices, which are apparently fine by commercial GIS format validation. With OGC validation, the duplicate vertices are repaired by removal. So the same (bad?) data can be served from a commercial geodatabase to a commercial ArcMap just fine, but if the same data served from PostgreSQL to ArcMap would create ArcMap errors. This was a bit frustrating to figure out and only occurred post ArcMap 10.5. Remember some organizations stay a version or a few behind the current release of ArcMap? If it ain't broke, don't fix it.

There was another error related to circular shapes in ArcMap. To fix this, a script was written to automate the repair. A grid system was created for each state to create smaller processing extents. Then each grid was looped through and processed using ArcPy. To eliminate circular shapes, a *densify* was performed, followed by a *simplify* and finally a *repair geometry*. Once these issues were addressed with pre-processing, everything ran smoothly between the database and ArcMap.

Another useful pre-processing step is joining tables. One example is NRCS soils data, which has an abundance of tables that can be linked to the map unit polygons. These tables provide all kinds of additional soil information and require joining tables for each county. Rather than having users repeatedly join tables after clipping data from the library, the tables were joined at the library level. This gave users everything they needed with the click of a button, ready to rock-and-roll with necessary attributes already in the extracted data.

10.8   Other Software Integration

Geoserver

Now what? We had a system that provided GIS layers to ArcMap and QGIS, but what about other users? Remember the second part of the problem, breaking down access barriers by extending GIS data access to non-GIS users? This system was a perfect GIS layer container; how could we expose the data contained in it even further? Was there a way Google Earth users could access the data loaded into the GIS Data Library? It turns out the answer is an astonishing "yes".

During this five year run and building upon the success of the library, it was discovered that GeoServer could be implemented as a front-end web interface. GeoServer is remarkable open source software that can do a lot of things, but one cool thing is accessing the GIS layers stored in the database and making them accessible via a web interface. GeoServer is really amazing software. It can take a number of GIS data formats and serve them over the Internet in different formats such as *Web Mapping Service (WMS)* or *Web Feature Service (WFS)*. It provides a web interface to preview, query and download data in various formats. GeoServer is a key piece of software that extends enterprise characteristics to the library.

Google Earth

As noted, one function of Geoserver is the ability to connect to a data source, such as a PostGIS table, then provide a variety of formats to download. The list is really long, but the common ones we know are available: Geopackage, Shapefile, and KML. When downloading a KML format, there are two format types. The first is a KML file that has the geometry stored in the file. This means that the KML file can be sent anywhere in the world since the data is contained in the KML file. This is great for viewing, but there are limits to the number of features that can go into a KML file and other things that make KML less than ideal for spatial data transfers.

The second KML format type is a network link to GeoServer, which makes the request to the back-end database. The main benefit is that the *KML network link* is very tiny since it acts like a pointer to the data. The main drawback is that the data source—a GIS Data Library—must be accessible. In scenarios where the data is confined to a *Wide Area Network (WAN)*, a KML network link sent outside the network will not work. However, for situations where organizations have a wide area network (WAN), this setup works remarkably well. This is a great solution to further expose the GIS Data Library to users who don't use GIS desktop software like ArcGIS or QGIS.

10.9 Success at the Enterprise Level

The GIS data library met enterprise standards and definitions. It was a GIS data storage and retrieval system, providing data to the entire organization. The library had immense success, covering 12 states with 6-20 GIS layers per state. There were over 100 state-wide GIS layers, centrally stored and distributed at blazing speeds. GIS users loved the library and showered the technology with praise and overwhelming support. ArcMap users had come to rely on this library for daily work needs.

The primary objective was met by using Postgresql and PostGIS to serve GIS vector basemap layers to ArcMap users. The ease of use with drag and drop .lyr' files in ArcMap was unbeatable. Being able to clip data from a centralized state-wide layer in a rapid fashion was one of the best GIS workflow enhancements ever created at this organization. The second objective was met with proof-of-concept work using GeoServer as the bridge between the library and Google Earth and web mapping. This capability provided GIS data access to non-GIS desktop users. By all accounts, this endeavor was an enormous success. Maybe the best story in this journey, considering the impact it has to an organization and the number of GIS users positively impacted.

In the next chapter, we'll go to a whole new level of theoretical data storage and distribution. A GIS data storage system that is globally distributed, secured by the users and has no owner. How can this be? We'll soon see.

11. Geoblockchain: A Theoretical Use Case for Blockchain and GIS Data

11.1 Something Different

The topics prior to this chapter were actual experiences with open source software: Linux, GIS analysis and modeling, web mapping and relational databases. This next chapter explores a theoretical use case for GIS data using a slow database technology called blockchain. After spending some time in the open source Bitcoin community, there was a curiosity to think about how this open source technology could be applied to a global GIS data storage and retrieval system. An Enterprise GIS was great for an organization, but what could be done for the global GIS community?

11.2 Just an Idea

Before anyone gets all upset, or starts making accusations, the example illustrated here is only theoretical; should it find its way into any development, or system for profit, it is a fraud. Rather than creating a new blockchain, invest your time into Bitcoin and figure out a way to build on it or use it. This is just for fun and to get people thinking in a different way.

One of the unfortunate adventures most people take down the Bitcoin rabbit hole is a diversion into altcoins. Altcoins are anything that is not Bitcoin. Buyer beware if you dip your toe in the waters of anything in the "crypto" space, this is why we focus on the blockchain aspects and avoid anything remotely related to financial instruments or "cryptocurrencies" in general.

11.3   Bitcoin

What is Bitcoin?

Bitcoin. A true technological marvel. What is Bitcoin? Good luck answering this question. Seriously. The attempt to explain Bitcoin has been done by many people, many times. Few times there is success in this explanation, but mostly not. It feels like we're going to have to explain this in order to relate blockchain to geospatial data. Bitcoin and blockchain are way beyond the scope of this book. It's beyond the author's scope of detailed understanding. Make no mistake about it, there is no claim of Bitcoin expertise. Just some late-to-the-game user experience with Bitcoin and testing the Lightning Network in the early days.

Looking at open source software projects, and there are a lot of them, it's hard to say which one is the biggest, best or even most impactful. At the end of any discussion, Linux seems to be the king of this title, but there is certainly another open source project worth considering: Bitcoin. Regardless of how it is categorized, it is undoubtedly the coolest open source project to ever be created (sorry Linux - no love lost).

The creator of Bitcoin, using a pseudonym of Satoshi Nakamoto, was a ge-nius. She or he had extensive knowledge about several distinct topics: cryp-tography, economics, computer science and peer-to-peer networking. Just like open source GIS software, when you're feeling good about yourself and all you know, jump into the technical details of Bitcoin to bring yourself back down to a humbled state.

In some ways, it's smarter to take the ambiguous road and say that Bitcoin is different things to different people. Some people see it as a currency or exchange of value while others see it as a long term investment. Some see it as something to bet with leverage (tales of Bitmex) and still others simply see it in a broader context of freedom. In truth, it's an open source project with great incentives. Not just an open source project, but an incredible example about what open source software can be with an enormous user base behind it, as well as developers who have the same passion. Bitcoin users and community are not unlike the users and community that surrounds open source geospatial technology.

A New Open Source Project Discovered

There is always something new to learn. It was 2017, a late arrival to the Bitcoin party, although depending on who you ask, "it's still early". Bitcoin had fallen from one of its famous parabolic highs in 2016 and somehow it was stumbled across in this journey. It had fallen almost 50% from its all time high. What is this asset?

The dive down the rabbit hole had occurred. It's interesting how some intriguing thing can grab one's attention and be all consuming. The pursuit of knowledge for Bitcoin was a new passion and exciting interest. When it was learned that this technology was open source and based on the idea of transparency, the curiosity could not be contained. Bitcoin did not rely on any third party to validate transactions or control them, it relied on the users and the code. The code became the judge, jury and executioner.

Sure, this freedom from large corporate financial entities sounds awesome, and in many ways it is, but it should be fully recognized that there's no one to turn to when a wallet is lost or a password is forgotten. Sure no one can stop you from making transactions, but there is also the fact that if you lose your keys and don't have a backup, you just lost all your bitcoin with no recourse to take. It's gone. The choice is certainly yours to make; you're free to choose if the risk outweighs the parabolic rewards.

Falling down the rabbit hole of Bitcoin was quite similar to falling down the rabbit hole of open source GIS software. There was so much to learn and with every turn there was a new side street to explore. A new idea or a new application someone had built. Mailing lists are really great for exchanging ideas, and in fact the creator of Bitcoin, under the pseudonym Satoshi Nakamoto, used mailing lists to communicate the development of Bitcoin. He or she is credited with being the creator of Bitcoin, others were involved like the famous Hal Finny who worked with Satoshi to make this dream a reality. Did we mention Satosihi vanished a few years later after Bitcoin took off and Hal is in a cryogenic state? These peculiar parts of the Bitcoin story are more than just interesting.[34]

Bitcoin and bitcoin

There is bitcoin (lowercase "b"), the coins, and Bitcoin (capital "B"), the network. The coin part is easier to understand, and by easier we mean still difficult. There are only going to be 21 million bitcoin created. One can purchase a fraction of a bitcoin, meaning X dollars can buy you 0.0X bitcoin, based on whatever the current exchange rate is.

The bitcoin is stored in something called a wallet, which is usually digital but can exist in paper format too. A wallet is just a collection of cryptographic

---

[34]https://www.alcor.org/library/alcor-member-profiles/hal-finney/

keys that access bitcoin at different addresses on the blockchain. The coins aren't even stored in the wallet. See, we're already introduced to puzzling complexity and this is just the user part, not the network or blockchain.

This chapter is mostly about the network, Bitcoin. The Bitcoin network is not so easy to describe, but in simple terms, it is a peer-to-peer network which can securely exchange bitcoin between users. Sort of, because bitcoin doesn't actually get exchanged, just the ownership of the bitcoin in the blockchain does. The network is unlike any other network, it is completely decentralized with no owner and has banking quality encryption security. It is only similar to other networks in the technical sense that it is a server software running on many machines—called nodes— using a specific port, 8333.

These Bitcoin nodes do not require a huge computing resource, aside from the initial process. They can be set up on Raspberry Pi Linux nodes and a medium sized hard drive. There is another layer on Bitcoin, called the Lightning Network, that can run on top of Bitcoin on the Pi node. The Lightning Network was designed to make instant bitcoin payments, since every 10 minutes the Bitcoin network validates transactions through block rewards.

Peer-to-Peer Networks

Peer-to-Peer networks are cool technology. Remember Napster? Napster was a centralized peer-to-peer file sharing network. Napster servers were needed to point users to the location of the files that lived on other users' computers. Napster itself did not hold the data, but it was facilitating the transfer of copyrighted material in files that were being transferred between users; namely mp3 files. The Napster server was taken down after a lawsuit. Once users couldnt access the index, their software didn't know where to look for data. This was the same problem with early digital cryptocurrencies; the central server was the single point of failure. If that could be solved, you could exchange something of value, money, with peers and verify the transaction.

Something technologically similar to Bitcoin, and even that is not quite right to say, is the peer-to-peer BitTorrent network. A network where users can share data with each other without a central authority. Users in a P2P network query the network for tracker information to learn about peer connectivitiy. Then the data bits start flying, or maybe flowing is more appropriate. Everyone in the swarm is a client and a server for some data.

With P2P technology, there's no central server to shut down. It was so good at what it did, there was no way to stop it. There was no company or kid to sue. There were websites hosting the torrent file, a file user's needed to join the P2P network and get data. Except now data was not just MP3 files, but high definition movie files too.

Long story short, media corporations use interesting measures, such as seeding an empty file with a name that looks like a copyrighted name. They get the user's IP address and other information to contact the user's Internet Service Provider (ISP). ISP tells customer to stop activities.

In the end, the P2P network for BitTorrent is still somewhat centralized, as it relies on a server(s) to provide the torrent file. Although better than the Napster configuration, it still has some issues to overcome to be like Bitcoin's decentralized network. If you find these things interesting, these are definitely some topics worth looking into.

Blockchain Database

RDBMSs are really good at securing and managing information. They are also high performance and able to be replicated across many instances to ensure constant uptime. This may not sound centralized, but in fact an authority figure could enter the system and make any desired changes. In a network of money, there can't be trust in an authority figure, only decentralized verification and concensus. All other digital currencies have been defeated from centralization.[35]

Blockchain is a special type of database that adds distributed security with every block transaction using cryptography. All the Bitcoin nodes update and track their database of coins and some mine for new blocks. Each block of transactions added to the database creates a cryptographic chain that can be verified and agreed upon by the network, thus users. This huge stack of intertwined blocks is the blockchain database.

The Bitcoin blockchain has heavily incenvitized security and stores the transactions and ownership of bitcoin. Sounds amazing and it is, but the penalty is that it's slow. Really slow. And this is by design because any big, sudden or unexpected outcome has the ability to impact many people's store of value. Slow and deliberate are not a bad thing and in some instances they are an actual benefit (e.g., government).

One of the many interesting challenges Bitcoin solves is the Byzantine General problem. This is a computer science problem where computers need to arrive at consensus, an agreement, but in a decentralized environment where parties (computers) are anonymous. This is already beyond the ntended scope and knowledge as everything has been overly simplified.

---

[35]https://bitcoin.org/en/bitcoin-paper

Satoshi solved this problem with Bitcoin and its blockchain, as more witnesses validate a transaction making it true, built on a long history of other verified transactions. In this system, bad actors are quickly ejected from the network and the worst they can do is double spend money.

What Makes Bitcoin Possible?

While there is an incredible amount of technology built into the Bitcoin network, what really makes it all work is the incentives. That's the secret. There are many things at play with Bitcoin that make it unique, but game theory is one of the more interesting topics. If you don't want to buy bitcoin, you can win some on the network if you guess the right number. This incentive was initially 50 bitcoin, then four years later halved to 25 bitcoin, and so on and so forth. With each halving, it's interesting to note the corresponding parabolic action not that long afterwards.

If we take a look at the price of bitcoin it seems that winning just one of these is worth the try. There are farms, enormous buildings with machines that are only trying to guess the new number to win some bitcoin. More computers hashing on the network increases security and the difficulty is adjusted as mining nodes come and go on the network. One way to try to win some as a little fish in the giant ocean is to join a pool where everyone participating shares the winnings.[36]

If no one wants the coins, then there would be no one hashing or guessing numbers on the network. If no one is hashing the network, there is no security. If there is no security, the coins are worthless. Using bitcoin as the incentive to secure the network is quite ingenious. It really is quite fascinating to read the founding email lists where these problems are being hashed out and solutions discussed.

Decentralization and Security

Once it was realized that Bitcoin was the only network with true decentralization and security like no other, the curiosity would not die. The idea was beyond intriguing that if half the world lost electricity, the other Bitcoin nodes would just keep communicating with each other and transacting. Almost like a virus that mutates and self-connects to move wherever the valid hosts are.

Creating a decentralized money network in theory is far different than getting one to work in the wild. The early days of Bitcoin are fascinating as Satoshi is

---

[36]Did we already mention "this is not financial advice"?

solving real-world problems as the network evolves. One example was spamming the network with cheap transactions, so a fee was implemented. There wasn't a problem that came up that Satoshi did not find a solution for. Simply incredible. There were others involved to their credit as well.

11.4 Those Other Coins

There are other coins similar to Bitcoin but all of the networks lack some key ingredient that Bitcoin has. There are many reasons Bitcoin can't be overtaken, and one reason is Bitcoin is the only one that works as advertised. How do we kmow this? Because it's open source software.

After having a basic understanding of Bitcoin it was quickly noticed that there were all these other coins. Apparently copycats take advantage of people through the complexity of concensus networks. The environment is filled with all kinds of wild characters who want your money through rigged money networks. There are minimal to no regulations on these networks, so all that can be said is be careful and be sure to do your research.

The only way these networks have validity is by an organic birth. For the birth of Bitcoin, there is now plenty of material out there to check out. It's definitely a cool open source project that is worth looking into. The best material is the origin story on the mailing lists—from Satoshi—where ideas are exchanged and Bitcoin is born.

11.5 The GeoSatoshi Idea

This chapter is really an homage to Satoshi. You may be thinking, "Bitcoin sounds cool, another great open source technology, but what does this have to do with open source GIS?" Like others, this idea, was sparked by Bitcoin. In the early days, Bitcoin used computer CPUs, then graphics cards, then dedicated machines to secure its network. GIS machines have great computing power and specialized graphics cards that are similar.

That "What if?" question. What if a blockchain, a geoblockchain, could hold geographic features while being secured by user's computers who are trying to earn coins to obtain GIS data?

While exploring mining with graphics cards, the thought about lots of GIS users with graphics cards existed. These users typically have machines that have upgraded CPU and graphics cards. Lots of users with GPU power could be used to secure a blockchain. The implementation of the open GIS Data Library was cool, but what if it could be extended to a global network of users with no owner? What if the network exchanged pieces of quality GIS data that had a built in user review process to ensure good data?

During non-work hours GIS workstations could mine coins. Users in all parts of the world could participate with the GIS machines they already have. Accessibility to all global users was paramount. There had to be incentives (Bitcoin's secret) to promote free data to help solve global problems. How can peer review be baked into this process and ensure good data? If it could work, it would solve all the user problems with GIS data: global one-stop GIS data shopping, easy searching, basic quality review measures and a standardized format.

After considering all this, a paper was developed and published on that professional networking site. The paper borrowed directly from the Satoshi Nakamoto white paper on Bitcoin. As silly as it sounds, a geoblockchain could easily be hyped in a Bitcoin bull market and sold as some next-gen GIS. Then, the infamous rug pull. So if you see it, let everyone know it's just for fun. Some of the altcoin ideas are totally absurd and this idea is right up there with tokenized car wash service.

Not long after this was published on that professional networking site, the publication was taken down. There were comments and questions from foreign interests that were not concerning until the epiphany about altcoins arrived; they're worthless. Some people realize that the characteristics of Bitcoin make it unique and quickly reject all other coins as fraud. When this realization occured, all altcoins were discarded. Even mining. The paper was shelved because initially it was a naive idea to spark interest in a global GIS data store, but it was concerning to think about the damage that could occur based on the internet inquiries.

If there's all this potential harm, why publish it in a book? For one thing, it open sources the idea if it ever did get out in the public. And the other reason, well, it's an interesting idea to share. Full credit to Satoshi for the inspiration and open white paper to borrow from. Maybe we can find a way to have a global GIS data store, without a blockchain. No one can build a better Bitcoin, but people can build on Bitcoin. Maybe it's you.

11.6 GeoSatoshi: A Blockchain for Geospatial Data

Abstract

A purely peer-to-peer open source ecosystem allowing for the aggregation, storage and distribution of geospatial data. Blockchain

technology provides part of the solution, but the main benefits of
this system are lost if a trusted third party is required to administer or control the system. Proposed for the geospatial community, is a digital commerce system built on blockchain technology, which encourages and promotes the creation of high quality
geospatial data. To incentivize high quality data, authors and re-
viewers are rewarded with the ecosystem's GeoSatoshi coin as
currency. Using blockchain technology solves several geospatial
data problems: accessibility, centralized ownership and global af-
fordability. Viewing the data is free to promote open access, but
downloading features requires the GeoSatoshi coin. A feature of
anonymity is proposed to eliminate coercive forces from intimi-
dating users and to promote data gathering in unorthodox, yet
ethical, ways. By way of the blockchain, data is not sent; it is
accessed with a key. The proposed system uses digital coins to
incentivize organizations to publish their proprietary geospatial
data to benefit the geospatial community at large.

Introduction

Geospatial data contains keys to unlock solutions to the environmental problems our world faces. It is paramount that this data be searchable and made
accessible to scientists and engineers around the globe. One clear problem is
that geospatial data has found its way into many corners of the Internet, in
both public and private space. Data continues to be built at an organizational
level, without an overarching technology to aggregate the data and connect
users.

Modern efforts to solve the data fragmentation problem work towards aggregating geospatial data into centralized spatial databases. Organizations invest
capital in people, hardware and software, and thus the resulting system and
data are only available internally to the investing organization. This capitalist
behavior keeps valuable and necessary geospatial data away from the larger
geospatial community. While enterprise spatial databases solve the data aggregation problem for organizations, these efforts fail the larger goal of allowing open access for the global geospatial community.

A system is needed to aggregate and distribute high-quality geospatial data to
the community based on cryptographic proof of trust, without the need for a
trusted party to centralize and own data. The project is expected to grow with
rapidity, given that there are decades of publicly available datasets ready to
store in the blockchain. More than a transaction system, an ecosystem is proposed to create a place for the geospatial community to exchange and create
quality data on a decentralized and scalable public storage system.

Ecosystem for GIS Users

Proposed is a trustless ecosystem with no central authority, that fosters a community for users to interact with each other based on cryptographic mathematics. Requests can be submitted to the community to build geospatial data in exchange for GeoSatoshi coin. This creates a decentralized and secure place to conduct an open exchange of coins for geospatial data. The proposed built-in, two layer quality control mechanism enforces data integrity on the blockchain. The use of the GeoSatoshi coin is also designed to incentivize organizations to sell their proprietary data to the community through the blockchain.

Geospatial Features

The primary asset in the geoblockchain is the geospatial feature. We define a geospatial feature as a point, line, or polygon having related attributes and values. The GeoJSON format is the proposed data format for storage on the blockchain. Features that belong to a GIS layer will use attribute values to associate the feature with the layer. Topological information is proposed to be stored with each feature, but topology will be enforced at the application level. Transacting at the feature level simplifies transactions and data sharing between users in the community. Features can be sent to users through the blockchain, without physically transmitting large data through archaic mechanisms like email.

Transactions

We define an electronic coin as a result of cryptographic work to secure transactions (Nakamoto, 2). Transactions occur at the feature level and require the GeoSatoshi coin. The blockchain solves the problem of double spending GeoSatoshi coins and double selling geospatial features. The decentralized nature of the blockchain provides no central authority for feature transactions. Transacting with features on the blockchain is proposed through a software application that functions as the project wallet and data exchanger.

Data submitted to the blockchain must be of the highest quality or the project will fail. In order to pursue quality, the project proposes a two-layer quality control system. The proposed system allows the community to help strengthen its own geoblockchain by participating in the enforcement of quality data. This proposition allows community members to directly add value to their coin by enforcing high-quality standards on data submissions to the blockchain.

Downloading a geospatial feature from the blockchain is a transaction requiring a GeoSatoshi coin. Uploading data to the blockchain is proposed through a two-step verification process. First, created data is submitted to a level 1 data pool. Then, another user of the system, who may or may not be known to the data creator, can peer review the data for level 1 approval. Chance may be introduced if the data creator does not have a peer selected to review their work. Approved level 1 data is submitted to the level 2 pool, where a randomly selected user is chosen for final quality control. Chance is the proposed measure to add spice to the ecosystem as well as anti-conspiracy protection for community users. Random user selections in the network are designed to oppose attackers from conspiring to game the system by creating and approving poor quality work for coin or the joy of destroying the project with erroneous data.

Transaction anonymity is a proposed option in the ecosystem to protect users. Users who review data for final submission cannot have anonymity because they are responsible for triggering payment to the user who created their data, the peer reviewer, and to themselves. The system can be gamed if anonymous poor-quality work is being created and submitted for other anonymous users to approve for coin payment.

Timestamp Server, Proof of Work and Network

"The timestamp proves that the data must have existed at that [published broadcast] time, obviously, in order to get the hash. Each timestamp includes the previous timestamp in its hash, forming a chain, with each additional timestamp reinforcing the ones before it." (Nakamoto, 2). This process builds the blockchain by locking blocks together with cryptographic hashes.

A distributed timestamp server can be defeated if an attacker amasses enough computing power to make the network follow the attacker's chain rather than the honest one. "To implement a distributed timestamp server on a peer-to-peer basis, we will need to use a proof-of-work system" (Nakamoto, 3).

The proof-of-work system keeps the nodes honest and cleverly defeats attackers. "If a majority of CPU power is controlled by honest nodes, the honest chain will grow the fastest and outpace any competing chains. To modify a past block, an attacker would have to redo the proof-of-work of the block and all blocks after it and then catch up with and surpass the work of the honest nodes." As shown in Nakamoto's calculations (Nakamoto, 6-7) the probability of an attacker creating an alternate chain of dishonest transactions grows with exponential difficulty.

Incentive

Proposed is an incentive system to reward the user community for quality work and to reward organizations for submitting their proprietary data to the blockchain for the user community to access. For the community, the system proposes a digital coin-based reward system for submitting quality geospatial features to the blockchain. The reward structure is weighted such that the creator, first level reviewer and final reviewer get 60%, 15% and 25% of the reward respectively. The reward proposition incentivizes private industry to publish their proprietary data to the blockchain in exchange for coin, thereby exposing more private data to the geospatial community.

Proof-of-work is the method to secure transactions and build the blockchain and generate community coins through GPU mining. GPU mining is the proposed proof-of-work, given that users of geospatial software typically use high-end workstations and graphics cards. This situation creates a perfect matchup for users to use their GPUs for blockchain proof-of-work. GPU mining is proposed over ASIC mining to allow more community members to be involved while minimizing user's capital expenses.

These two proposed incentive methods provide a solution for users to earn coin without investing a lot of monetary capital. The incentives are designed to be fair to all members of the community regardless of their socioeconomic status. The proposed system will not solve the data inaccessibility problem if users cannot afford to monetarily access the data. If at any point the project is found to be restricting data from the community, the project should be redesigned or abandoned as the main purpose of the project is to put geospatial data in the hands of those who can improve lives or our planet.

Organizational Node Hosting

Organizations often have GIS Professionals clustered geographically, with GIS users spread out globally. Enterprise configurations for GIS typically centralize data in a spatial database for client software to access. Historically, organizations have invested a lot of capital to build and acquire geospatial data. Organizations, both public and private, play an enormous role in the development of geospatial data. Organizations should be seen as an ally to the project.

When an organization hosts a node, they are a peer node on the network. Organizations have no authority over the system and the only control they have is turning the node on or off. If they misbehave, the community can exclude the node. Organizations provide a benefit to the community by providing resources to strengthen the network through the node they host.

The benefit organizations receive is faster access to high-quality data without the expense of maintaining a data server. It can take a long time to download geographic data because of their potentially enormous size. Local node hosting can be extremely beneficial to organizations in parts of the globe with slow Internet access.

Simplified Payment Verification, Combining and Splitting Value, Privacy & Hostile Takeovers

The proposed ecosystem uses the same features from Nakamoto's success: simplified payment verification, combining and splitting value and privacy. Simplified payment verification is applicable for checking the chain without running a full node (Nakamoto, 5). It is achieved through the use of headers. Rather than making the system handle every cent in a transfer, value can be split and combined (Nakamoto, 5). Privacy works by allowing everyone to see every transaction, but not know who is attached to the address in the transaction (Nakamoto, 6). The proposed ecosystem is intended to build a community, so privacy should only be used where users need to protect themselves. Nakamoto's white paper calculates the odds of an attacker taking over the network using the proposed infrastructure (Nakamoto, 5-8). The best an attacker can do is take over the network. This would require the attacker to generate an alternate chain faster than the honest chain (Nakamoto, 6).

Conclusion

To move geospatial data storage to the future, a decentralized, secure, community-driven global geospatial data storage system is proposed. Centralized spatial databases have served the community well and will still play a role in the near future. Blockchain technology can scale geospatial storage globally and securely, with no central authority. Blockchain technology solves many geospatatial storage problems: data aggregation, public storage and access, private/public access toggle, feature version tracking, transmitting data to users, transparency and capital expense for special data storage and backups.

Embedded in this ecosystem is the proposition of a geospatial data exchange, with the primary beneficiary being the users of the community. The system proposes to provide a mechanism where members can use GeoSatoshi coin to have community members build requested data. The GeoSatoshi coin can also be used as an incentive to encourage organizations to publish their proprietary data to the geoblockchain for community use. A two-layer quality control system is proposed to promote high-quality in the geoblockchain.

The GIS community has a very impactful presence in the open source community. There are a large number of high quality open source software projects that work with geospatial data. GIS software is available for free across the globe, but geospatial data can be hard to locate or broker. Many open source geospatial tools and components are already in existence; the blockchain is the revolutionary component that has not yet been integrated.

References

Bitcoin: A Peer-to-Peer Electronic Cash System; Satoshi Nakamoto; `https://bitcoin.org/en/bitcoin-paper`

12. The Big GIS Secret

12.1 Big Secret?

Sounds like click bait for sure. What could be a secret at this point? We went through a good deal of technical software, applications and roles. We also saw how open source database systems can provide enterprise GIS data services through a library and touched on the needs it fulfills in an organization. The secret has more to do with the overall success of GIS in an organization rather than dazzling GIS analysis, modeling or systems.

This secret is not so much a secret with people who understand GIS data challenges and how they can exist in different network situations; ranging from in an office (LAN), to in an organization (WAN) and to the Internet (WWW). A single GIS user providing GIS desktop services may not encounter any issues providing maps and data files, but once we get beyond the desktop user we see that there is a bigger need to be addressed. Some organizations may implement these best practices just for the technology benefits, but as GIS users, we have a bigger objective in mind.

12.2 Putting it All Together

It wasn't until later in the journey that the puzzle pieces came together to form the overall picture about how GIS best fits into an organization. In the early years, there is protection of the job, knowledge and capabilities. There is perceived value in feeling like people need you for some task to be done with GIS software and data.

This exclusivity is completely normal during the early years of a journey, but at some point it becomes a problem. Restricting information and knowledge is rarely, if ever, a good thing. The idea that there is value in isolating data is eventually proven false. It took a while to realize and accept this fact, but eventually the secret to success with GIS was revealed.

12.3   Top Five GIS Mistakes

Before we get to the big secret to GIS success, let's take a look at the top five GIS mistakes made by users and organizations. These observations have been made over the course of this journey.

#5 - "GIS System"

Can't recall how many times this has been seen or heard, "GIS System". Such as, "Can't we just press the easy button in the GIS System?". A bit irritating. It's Geographic Information System people, putting the word "system" after GIS is Geographic Information System system. Let's use the acronym correctly, especially if it's your profession.

#4 - Mis-Labeled Analysts

This one is more of a function of the organization than the worker, assigning the wrong title to a GIS user. Much less problematic today, but in the early days, mistitled analysts were frequently encountered. The worker may have been someone who was assigned the title by an ignorant manager. GIS has become much more embedded and accepted in organizations with titles that match capabilities, but in the early days, even a person running a script was an "analyst".

#3 Equating GIS to Commercial Software

It's very encouraging to see that this too is fading in the US, with much thanks to the popularity of QGIS. There are some parallels that can be noticed between the Windows operating system and ArcMap. In the 2000s, there weren't really other user desktop operating systems to use besides Windows. One statistic was that Windows was on 95% of desktops. There were some Apple computers in use, but not as common as today. When people say "computer" there was, and still is to some degree today, an immediate association with the Windows operating system.

The same thing can be said about GIS being associated with a particular commercial software company, specifically in the US. A GIS is a collection of spatial geometry functions and SQL queries, not a specific software brand name. Likewise, Linux and its variants are well established in the server computing world, though Windows gets more of the consumer press.

#2 Sloppy Data: Slivers and Gaps

One of the biggest "mistakes" noticed in GIS data over the years relates to sloppy data. Sloppy data in this context is inaccurate digitizing resulting in slivers that have a gap between polygons or an overlap. Sometimes this data appears fine visually because the inaccuracies are so tiny. Most sloppy GIS work is found in the details, usually discovered in table summaries or a dissolve function.

There are tools which are designed to rectify slivers and bad topology, but in some cases the data has to be manually repaired if a GIS tool is unacceptable. The concept and use of topology seems to have vanished in GIS and has been replaced by hack-and-merge GIS techniques; such as cutting a larger polygon into smaller pieces.

And the #1 mistake in GIS goes to... you guessed it. Drum roll please...

#1 Commercial GIS Software

The number one mistake in GIS is commercial software. Not surprising, considering that previous chapters could easily be doubled in content or be books unto themselves; and that's just one personal journey. There are at least 350,000 users of commercial GIS who have unpleasant tales to tell.[37] We talked about the commercial software crimes committed against GIS users and more broadly, the GIS field. The attempts to commercialize and proprietate open data formats is at the top of the list; it's science data afterall. If the only benefit from a special proprietary GIS data format is gaining more profits, then we must get rid of that format.

The "for profit" actions of commercial GIS companies discourage the interchanging of data between systems; and no, 30+ year old shapefiles are not an acceptable geographic data format in the 21st century. Commercial GIS practices that corner the market with vendor lock-in ultimately harm the whole GIS community. Closing off this market decades ago, they have no real competitor except open source GIS.

We already hashed out the first decade of ArcMap woes in chapter four, but to be fair, we should tie this into a more recent time so as to not let it be perceived the complaints were just from 15-20 years ago. Around 2018 in the ArcMap software, a neat bug had been introduced. Yes, this was a brand new bug with one of those fearful, rather than delightfully anticipated, upgrades. We started

---

[37] https://www.gpsworld.com/gis-users-come-from-every-field/

noticing a directory with the name "1" appearing in all the workspaces with a .mxd file. The "1" directories are empty and just temporary files that ArcMap creates.

This had to be present in hundreds to thousands of project directories. Tech support said the solution was to reset the .mxd' template. Uggh, this means all the user placed buttons, toolbars, extensions, and window docking get reset and would need to be set up again. Just one example of the continued bugs throughout the 20-year lifespan of a commercial GIS software.

Linked with this commercial GIS software disappointment is the diminished technical support. Tier one support offers not much more than reading the help files to the user. Tier two is harder to get, but is more knowledgeable. Oftentimes users are directed to online forums and community help.

This could turn into a lengthy list of complaints, but we went there and it's old history. Feel free to do your own research and see for yourself. Better yet, don't even waste your time with it—unless you find it intriguing—and just read some instructional books on QGIS or GRASS GIS. It's much better to spend time learning about stuff that works, than learning about stuff that doesn't work.

Commercial software:

- Unjustifiably expensive
- Favors profits over quality and user needs
- Seeks vendor lock-in
- Low quality software
- Low quality tech support

Can't forget about the creeping annual fees to maintain this software as well, that's the real silent reduction of profits over time. These factors, coupled with the propaganda that commercial software is the "best and only" GIS in the market, make commercial software the number one mistake in the GIS space.

12.4   The Secret to GIS Success

Unfortunately, the secret to GIS isn't being a great desktop modeler or analyst, it's an idea bigger than oneself. Leaders in GIS have increasingly supported high integration of GIS technology within organizations. Empowering other users with GIS software, services and data is the real secret to GIS success. Fetching data for people may meet a request, but teaching users about GIS will empower them for a lifetime. Of course that's based on the "teach a person to fish" adage, but it goes deeper than just teaching others what buttons to press. Users benefit the most when they learn about GIS as a technology and the foundational components that go along with it.

Share the GIS

How to succeed with GIS? It's an interesting question because it causes the reader to think outside themself for the best answer. Success is usually defined around oneself, not others. Sure we can have our own success with GIS software, but that's kind of a given, isn't it? People usually act in their own self-interests. We see something, we pursue it, we learn it and we use it. Your own growth and knowledge expansion will be obtained roughly proportional to the amount of effort put into it. Real success with GIS, really *for* GIS, is making GIS software grow into an everyday commonly used piece of software; expanding its reach and user base. This can be accomplished in different ways.

One of the most straightforward examples we can use is the origin of the amazing QGIS software. Gary Sherman is the father of QGIS and has done an incredible thing creating this software.[38] He's undoubtedly done many other incredible things we don't know about. We all owe him a word of thanks. He gave us this cool software to start using and then the community built on it. Just how much did the community build on it? Impressive statistics like expanding the code base by 2,000 times over ten years. They turned it into something completely on level with commercial GIS software, taking it far beyond what one person has the time to do. This is just one example, but in the field of open source GIS the idea is the same: create and share.

> Thank Gary and learn to develop on QGIS by buying his book: PyQGIS Programmer's Guide 3 (locatepress.com/book/ppg3)

How to succeed with GIS was an intriguing question, especially after all the years invested and the GIS technology experienced. At this time, there was already a progressive mindset that open GIS data access should be a standard. In this way, data access was brought to users through an open GIS data library. It was clear that widespread GIS use has a multiplier effect, because it's not one person doing GIS work, it's 5 or 30 or 100 people doing GIS work in a uniform way. The 'uniform way' part is essential if there is to be any order and overall organizational benefit from the software.

---

[38]https://spatialgalaxy.net/2022/02/18/qgis-then-and-now/

It was during the development of the open source GIS data library that all remnant beliefs about keeping GIS a mystery and job security drifted away. Years of expertise and accomplishments also build the confidence needed to stop looking inward and start looking outward to see how GIS can be used in a bigger way and by more people. Below is the specific section of a blog post that crystalised the idea and turned the opinion into an unshaken belief.

How to Succeed in GIS Blog Post

"Here are a couple of quick observations I would share using GIS efforts as a basis. First, the strongest enterprise efforts are not defined as one or two individuals that do all the work for every department. Put the tools in the hands of everyone and let them use technology as part of their work. You won't be put out of a job. You'll find you are actually in more demand. Second, lead wherever possible with applications that improve productivity and efficiency. You will see a quicker return on investment through a digital transformation. Third, document your return on investment. Doing so is a strong accountability practice. It also aids in reminding personnel at all levels why you are investing in technology. And it's great to have handy when budgets are tight."[39]

This was incredibly concise and correct. The key part was calling out the old school way of using GIS software and users in an organization. In the early days, it was cost prohibitive for everyone to have GIS software, so a few users ended up doing all the GIS work. That's how this journey started, one user fetching GIS data that later evolved into an open enterprise GIS database.

It's logical to follow that by putting more GIS software into the hands of workers, there would be an increase in usage and thus demand. Expanding the GIS user base grows GIS demand, however, this growth cannot be with rogue actors and isolated buckets of data on network servers or local workstations. There needs to be an enterprise approach with, at a minimum, guidelines for using base data layers, template maps and where newly downloaded data should be stored.

This basic idea of "Share the GIS" fits the natural progression of technology within a broader context of societal use. Newer generations are immediately exposed to technology. Toddlers can easily navigate tablets and then later kids are modifying Minecraft code or adding exploits to Roblox. What used to be a small community and niche for computer nerds has gotten mainstream

---

[39]How to Succeed in GIS Without Really Trying, Christopher Thomas, https://www.govloop.com/community/blog/succeed-gis-without-really-trying/

to where kids can really get involved at a technical level, even if they don't understand the details.

Software use has become commonplace. Technology is everywhere and the exposure kids get is far more than 40 years ago when it was extremely rare to be tinkering with computers. The technology evolution is a natural development and GIS falls right in line with it. GIS is taught far more in university and GIS software is far more common in the workplace than it was 25 years ago.

Encourage GIS Use

The message is simple: let them play. So, GIS is here and everywhere. Now what? One of the key components to succeeding with GIS is to have users engaged with the technology. Experience shows that people become upset when they can't play with GIS, so embrace their desires and let them play. Growth of GIS users should be encouraged as we want GIS software to be as seamless as using a word processor or email. We already see this happening in some organizations where enterprise GIS deployments are in full usage.

If you're the GIS expert, don't be afraid to encourage and allow other users to perform supervised activities that can contribute to a larger GIS effort. Once they become a trusted user, concerns fade away. As people expand their skills, applied GIS technology grows within the organization. Build relationships with people as much as one does with the software; especially if you're an introvert. It's good to have a wide base of people who can use GIS software because when the knowledge is spread out, there are fewer bottlenecks to meet simple data needs. There is no doubt; you definitely don't want to be the bottleneck when there is a boatload of GIS work that needs done yesterday.

One way to ensure a consistent baseline for GIS users is to use systems that confine user behavior, or even better yet, incentivize them to do a desired action. Bitcoin is owed some thanks for this lesson in incentives. GIS experts can use GIS infrastructure to create collaborative systems that guide users into consistent data use and outputs. One example is a GIS data library, ripe with GIS data fruits just dangling and waiting to be plucked by hungry users. GIS users love easy to access and cataloged GIS data layers. They just can't resist.

12.5   GIS in Organizations: Technology & Users

How to View GIS Technology

The answer to the question, "How do I succeed with GIS?" contains two components: software and users. The technology is just software and someone needs to know how to administer it and be an expert using it. GIS is just like other software an organization will need to use for its operations. The technology has several routes to reach end users: desktop GIS, web mapping, Google Earth and open GIS RDBMS layers. The software technology is the easy part, but the GIS data requires some consideration; file-based format or an RDBMS solution. The goal of the GIS technology is to allow a broad range of users the capability to edit and access GIS data; on second thought, at least just access it.

Make Use of Non-Dedicated GIS Users

The non-dedicated GIS users are the ones that can have the greatest impact in an organization with GIS. Within an organization, there are non-GIS professionals who use GIS as a tool: planners, scientists, engineers and more. There's not much to consider with regards to GIS technology for these users, because they use GIS software in conjunction with their primary field of expertise. These users are usually in non-GIS departments so there is no issue with where to place them. The main concern is ensuring that they have access to standardized GIS software and data. They know what they need to do for their job, and if things get too wild, hopefully there is a GIS expert or professional available to consult.

What is important from the organization's perspective, is to be sure to adopt the modern approach to a well integrated and flexible GIS user. The old idea of cramming a ton of work through a few users is dead, as new users can wield GIS software with ease and a smile. This practice disperses the organization's GIS resources and reduces the number of dedicated GIS software users sitting idle waiting to get cut. Higher GIS integration helps increase job stability for the worker and reduce turnover for the organization by balancing staffing needs appropriately.

Dedicated GIS Users: Experts, Advocates and Mentors

The biggest question is what role a GIS expert will play in an organization, because there are several roles that can be fulfilled. This journey traversed GIS as a solo GIS user, leader and worker of a GIS production team, analyst in a GIS department within a corporate services department and then a GIS analyst in a natural resources service line, which is like a broad department. The use of GIS has been observed over a long period of time, under different management practices, but unfortunately they were all the variety where a few GIS users do all the GIS work.

We know that the success of GIS is not by one user, but many. This opens a whole new line of questioning about how and where users fit in an organization. Should GIS users be non-GIS professionals who can use GIS software or should users be dedicated GIS professionals with focused expertise in GIS technology? Should GIS users be placed in a non-GIS department or should they be in their own GIS department? Should they be a part of corporate services like IT workers? It all depends on the size of the organization and how much it relies on GIS data for its operations.

GIS Expert Roles

In order to accommodate this progressive approach to GIS integration, the GIS expert ultimately has three roles: support users, develop the technology and provide services as a specialist or analyst. In supporting users, their role becomes similar to IT professionals: cultivating an environment that allows GIS users to grow. When viewed as an expert resource, GIS professionals can be utilized as educators and advocates of GIS technology. This involves seeding the ground with GIS data in an enterprise database, and using no-cost GIS desktop and web software to harvest the data.

All of this involves staying current with geospatial technology and trends. Working with GIS users, GIS experts can identify their needs and develop tools. When a task becomes repetitive, the experts can develop code to automate the task for users, smoothing out the work flow and increasing efficiency. They also promote and develop geospatial technology by deploying web mapping, mobile data collection or enterprise databases. One GIS power user can lead and guide many users to a consistent, happy and productive outcome using GIS. Without this GIS lead guidance, GIS will become isolated in the organization as users collect their own data and develop their own standards.

The obvious role for the GIS expert is to conduct high-end analysis or modeling. These applications are usually developed over years of experience and exposure to different problems and the tools that solve them. The expert GIS user has both breadth and depth of GIS technology. It is expected that if they don't know how to do something that they will be able to research and figure out how to do it. If the analytical role, and not the other two, is the only one focused on, it's entirely likely an organization will end up with too many dedicated GIS users and not enough work.

Large organizations require dedicated staff to fulfill roles in an enterprise GIS. With a GIS subject matter expert, smaller organizations can leverage the technology for similar outcomes, using fewer people in hybrid roles. Hybrid roles are a great way to get things going if the organization is not large enough to accommodate dedicated staff. For example, IT staff are perfectly well equipped to set up and administer GIS web mapping and databases, so be sure to have those doughnuts ready. Most organizations have a basic database person and if they can learn to work in an open source GIS RDBMS with a spatial component, then this small hybrid team is complete with the addition of a GIS expert. Small services can be deployed and scaled up with more resources and staff as needed.

GIS Staff Placement Within an Organization

One of the biggest questions seen thus far, kind of humorously to a GIS user, is where do we put these GIS employees? Just one of these weirdos can go anywhere, but if there is a collection of them, they should be kept together. For one thing, you won't want to hear their geobabble. The nerdy geotalk will drive you nuts. The topic of a GIS manager is a whole other issue, but we'll assume there is no need for a manager, workers can self-manage. GIS is a big field and there are lots of different things to consider: standardizing templates, data formats, data filing systems, map outputs, symbology and base data to name a few. When there is just one GIS person, they set those standards, but when there's a cluster of these mapping maniacs, standards become necessary.

Some of the roles for GIS professionals involved are similar to IT, so placing the crew in corporate services or alongside an IT group may also be appropriate. Situating them as a corporate resource helps signal to the group, and the organization, that they are available to help users, not just do work. If they aren't seen as a resource that can be accessed, users will go off and do their own thing. Building this strong connection between co-workers is part of the secret to success with GIS.

Integrate Users or Technology?

Both would be the easy answer. There once was a question about what to do with dedicated GIS employees. The question was off in some way though. The question was, "What are the plans to get the GIS team better integrated throughout the entire organization, regardless of department?". Integrate the users better throughout the entire company? Good question. Drinks and food are always a solid go-to. It was viewed as an odd question because do we look to integrate GIS users or GIS technology?

This question best illustrates the disconnect in management's view of GIS. In consulting, worker's time is charged to clients. So everything is viewed from the perspective of "How do we bill that person more?". This kind of question focuses on profiting from the employee, not the technology. In the early days of GIS, it made sense to charge for a special operator with special expensive software. As technology evolved though, it made less sense and the old school thought missed out on opportunities to bill for GIS services rather than people's time. The goal is not to get the dedicated GIS users (a group) better integrated throughout the entire company. If this question is being asked about integrating GIS workers, then the GIS group of experts are too big for the organization's GIS needs.

It's an abstract goal that, minus food and drink, won't be achieved very well. We know how to integrate people better, but if we're looking to increase the use of a role, the most obvious way is to create demand. One clear analogy is IT workers. No one is looking to integrate them, but their technical knowledge about software and hardware are things that are in demand. The IT team integration happens naturally because of the demand. No one needs to look to integrate IT staff, they are integrated out of a business necessity. In fact, the questioning is kind of stunning because that's how all roles work in an organization. The How To Succeed with GIS blog didn't suggest to put users in the hands of everyone (eeek), no, put the technology in the hands of everyone.

Benefits of Distributed GIS Software & Users

We're now looking at a well distributed system of GIS users and software, with at least one-to-many knowledgeable dedicated GIS users in the mix as guides. The configuration isn't rigid. This is just suggesting that the greatest benefits come when GIS technology is well integrated in an organization and not confined to a few people doing all the GIS tasks. When the technology and basic tasks are well distributed, the organization really has the potential to thrive.

Some benefits seen from integrated GIS technology are:

- Minimal idle GIS staff
- GIS skills expanded in organization by expert mentors
- GIS technology exposed to different users - new ideas
- Reduced work-flow bottlenecks
- Greater project data interaction with workers
- Software matched to user needs

- If must be in commercial trap, integration minimizes financial impact

- Access to GIS subject matter experts
- Standardized software, data and map templates

12.6 Enterprise GIS

If we take this whole mash-up of desktop, web, RDBMS, mobile and link it all together what do we get? A bunch of software pieces that each relate to GIS, some directly and others indirectly. What we end up with is a collection of software that works together as a system, servicing all kinds of users in an organization.

There is GIS in an enterprise and then there is Enterprise GIS. To start with, what is an Enterprise GIS? In the simplest form, it's a system—a collection of software—that stores and distributes GIS data and services throughout an organization. In other words, GIS database servers, desktop workstations, web software, mobile data collection; everything we've talked about so far. Can you build an open source Enterprise GIS? Well, you made it this far, what do you think? Bet you have the right answer.

Enterprise GIS: The Commercial Version

The "How to Succeed with GIS" question is really talking about Enterprise GIS, but Enterprise GIS has a commercial GIS connotation; at least in every situation encountered thus far. We've discussed open source and commercial software, so what is enterprise GIS using commercial software? It's a full collection of vendor specific software and proprietary data formats, wait a sec, this is starting to sound familiar, that makes it impossible for users to switch to another vendor. No, hold-on, that's vendor lock-in. Well anyways, it turns out they're the exact same thing. A collection of [commercial and proprietary] geospatial software that stores and distributes information to an organization.

> Spatial Data Infrastructure (SDI) is also commonly, if incorrectly, used to describe the set of components used in Enterprise GIS.

Enterprise GIS is a term that, as far as we know, was created by the commercial GIS software company. Quite obviously, they sell a collection of software that services an organization or enterprise. Enterprise GIS, for the commercial

GIS software company, is using all of *their* commercial software to store and distribute GIS data and services to an entire organization. Enterprise GIS in commercial software is the exact definition of lock-in. Vendor lock-in didn't sound so customer friendly, so here we are with Enterprise GIS. No one would change systems once all the components have been bought, configured and are in "working" order. All that's left to do is perpetuate an infinite stream of annual payments.

Remember the mess involved with trying to get an open database format to work with ArcMap? It took a whole chapter. The commercial GIS company designs their software to work best with their own software. Sure, they advertise these interoperable formats that can work with open source, but the reality is, their marketing team is going to propose all of their commercial products, not other options. They're goal is to jail the organization with their software. How much would we guess organizations are willing to pay to be put in jail? Hundreds of thousands of dollars per organization is not a wild estimate.

What About an Open Source Enterprise GIS?

A collection of software that provides open GIS data storage and access to an organization. It's the same definition, but in the context of open source, it takes on a whole new form. There's no vendor specific brand. Enterprise GIS is an abstract concept but is made real through software components. At the core is an RDBMS, desktop analysis software, desktop mapping software, web delivery services and mobile data collection. Can some of these pieces be missing? Sure, but the definition gets more abstract. If there had to be one defining piece for an enterprise GIS, it's the RDBMS. The RDBMS is a key piece of technology that provides single container access to all users in an organization, while supporting advanced capabilities like multi-user access and editing.

We identified generic software components for an open source enterprise GIS, can you name some of the specific open source software? PostgreSQL and PostGIS are an excellent choice for the core RDBMS technology. GRASS GIS is a top-rated GIS desktop analysis software, which also connects to Postgresql/-PostGIS. QGIS is excellent desktop mapping, editing and analysis software and can also easily connect up to PostgreSQL and PostGIS. Web services can be set up with Geoserver, which can also read from PostgreSQL databases; noticing a common theme? GeoNode is an project-driven GIS application that helps organize and create project specific map groups and has searchability, of course plugged into, you guessed it, PostgreSQL. Now we just need a way to collect some data in the field and ODK is right there waiting. ODK was used a long time ago and should be investigated, but historically it was a great solution. Many now use Mergin Maps for mobile integration.[40]

So to answer the question, yes, of course we can build an open source Enterprise GIS. The cool thing is you can pick your own favorite open source software to plug into the enterprise GIS software collection. You like some other database or operating system, go for it. There is no such thing as vendor lock-in in the open source world, so plug and play until your heart's content. Oh, and no hard feelings to any open source GIS project not mentioned, they all are worth considering as each has its own benefits and community.

12.7   Knowing is Half the Battle

Most organizations implement the secret to GIS success, even if never having read or heard about the concept. Especially large organizations, where this kind of GIS technology integration is managed by IT workers who recognize the value in Enterprise GIS strategies. For other organizations, GIS software has been acquired at different points in time in different locations. There may be one or many licenses and lots of data spread out over many network systems. These organizations are ready to reap the benefits of open source Enterprise GIS.

---

[40]https://github.com/MerginMaps

13. Can't Beat Passion with Payments

13.1   Why Limit Yourself?

If it's not clear by now, then the objective of this book has failed. Commercial GIS software has a closed format and limited options that makes it hard to work with other software. Not only that, but the limitations imposed by commercial software limit user's thinking and capabilities. The software costs are outrageous with the initial purchase and even more to maintain over a long period of time with annual fees. On top of all that, the quality of the software isn't that spectacular.

What are the benefits of commercial GIS software? Are they worth the initial cost and annual fees? Is vendor lock-in ever worth it? Freedom is just a choice and nearly all choices are surrendered once an organization finds itself locked in a vendor software jail. Buying commercial software is not an investment, because every year more money has to be added to the asset just to keep its value. It's more like renting different software versions in perpetuity. What could organizations do by truly investing hundreds of thousands of dollars into something else? Just about anything except commercial software would be a better choice.

It's clearly detrimental to an organization to achieve vendor lock-in. There has been a lot of time and experience to consider these situations and one saying that keeps repeating in every instance is: "you can't beat passion with payments". Passion in computer projects is really hard to beat, if not impossible. Bitcoin is a great example, but for those involved in the open source GIS revolution, it's clear that this crowd has just as much passion if not more. Payments can be made, but the commercial GIS giant will just lumber forward without your consideration or regard. Watching commercial GIS software failures while catching the open source passion for GIS projects proved one thing to be true: you can't beat passion with payments. Not by a long shot.

13.2   Proceed With Caution

Remember that GIS is just software. Dedicating your career to GIS means you are dedicating your career to software. Sure there are concepts and specialization, but ultimately it's pressing buttons or typing in code. Be sure to learn some underlying science, because after 10,000 hours, where will your expertise be? It will probably turn out just fine, but do apply some caution in putting your whole career into the hands of software. Software changes, so be prepared to adapt at least a few times over your career. There was a danger in only being taught commercial GIS in university because when commercial GIS got so miserable, all of GIS was almost abandoned, not just a commercial brand.

With GIS, the users that excel the most are the ones that can use it beyond pressing a sequence of buttons to get some results. They stand to gain the most by understanding the data, the processes, and how it can be used to model instances. When GIS becomes as ubiquitous as email, be sure you have a special skill, or else you'll be out of a job when any entry level worker can make a map or do an overlay. Everyone can type a word document, there's no more document specialists in the way they were.

Keep growing and learning your skills and this will become a miniscule concern. Don't be stagnant. If you love GIS, keep growing by learning as much about GIS as you can. If you don't care for GIS that much, find a field you love and then learn as much GIS as you need to do what you love.

13.3   Keep Your Chin Up

Things won't always go as planned, but keep going forward. It's a continuous learning experience, where one experience builds into the next. GIS professionals are in the business of using GIS technology to empower organizations to jump to the level. We do more than create maps, or even modeling and analysis. We have the ability to empower an organization with geographic intelligence in the best and most efficient way. If you find that the organization you work for exhibits ambivalent responses to your technological marvels, leave. It's just a matter of time before they eject you or they are totally eradicated in the marketplace by organizations that embrace this technology.

13.4   The Future of GIS

What is the future of GIS technology? Hard to make predictions about the future because the flying Jetsons cars haven't even arrived yet. Not much has changed in GIS. It's kind of like the automobile in that the basic parts are still the same, but a lot of new technology has changed the automobile. In the same way, the web has changed GIS. Paper maps were commonplace and now we use web maps to create an interactive map in a browser or capture maps into a digital format such as a PDF.

There is a future with augmented reality and GIS. The idea of wearing something that can give peripheral information may be a useful thing. We never imagined being glued to phones and maybe the next step is digital glasses. Either way, there is a location and a database waiting for this application. Maybe one day we return to walking down the streets to shop, with digital pointers leading us to a sale in a small shop. It would be nice to see small business and community return. Virtual reality is another place where worlds will need to be modeled and adventures will be had. Maybe your next vacation is a family trip to the Grand Canyon in a VR headset.

Big data is a growing field as our devices constantly collect information about us. Since we travel with them, they know our routines and travel places as well. Today we have lots of information and since GIS is an information system, there is a perfect match. A good deal of the data has a geographic component. Big data is everywhere and this is very good news for GIS jobs. We haven't nearly mapped all the stuff on Earth yet and we keep adding. There is a lot of opportunity to use GIS to filter, sort, and display these large volumes of data collected over time. There's a projected 15% year-over-year growth with GIS, so it's still chugging along and will for the foreseeable future.

The future of open source GIS has never looked better. Geospatial open source projects have gained a lot of traction and users and have rapidly accelerated these projects. Open source solutions are much more accepted today and the technology easily competes at a commercial level. There are more jobs posted asking for open source software experience, namely QGIS. Not enough things can be said about the communities surrounding open source projects, something that makes open source GIS unstoppable. The price tag of "free" helps get open source software into users hands, and the quality and community makes them stay. There is no doubt, the future looks great for open source GIS.

The future of open source GIS is particularly interesting because it is like Bitcoin, having users who are fiercely loyal to the underlying philosophy of the software project. These people are nuts; that's meant in a good way, meaning passionate and defensive. Similar parallels exist with open source GIS because of the users. The user community cannot be emphasized enough in open source. It's global, decentralized, and very passionate. It's these characteristics that will raise open source GIS above commercial GIS, particularly in the US. You can't kill Bitcoin anymore than you can kill open source GIS projects.

13.5   Is There a GIS Professional?

We asked several related questions throughout this journey, two of them are: Are there GIS Professionals and is there a GIS Profession? We gave plenty of examples and looked at a bunch of different scenarios. Well, after all this, what do you think? Is there a GIS Professional or just a GIS software user? How should a GIS Professional be defined and how do they earn the label? Is GIS a Profession or an applied spatial technology? Maybe it's both? In the end it may matter little or none, the important thing is that you can explain your own answer with some good rationale. Not sure there really is a right answer, but we've all got opinions.

The opinion formed by the end of this journey is that there is a specialized knowledge base required to use GIS technology correctly. But we're talking deeper than making a map, we're talking about the science behind it: projections, measurements, shapes, data, and analytical functions. The data can't be emphasized enough. The database component falls under computer science, which we know is a discipline of its own. It may not be special software, but it certainly is spatial.

Under it all, GIS is rooted in science. Since there are disciplines within the science branch, and GIS is one of them, GIS is a discipline or profession within this area. And since there can be a discipline within the science umbrella for GIS, with dedicated users attached to it as a career, therefore there are GIS Professionals. There, are you happy? We finally answered this ridiculous question.

13.6   Go Get Some!

The GIS field is large, really large, too large for anyone to know everything. There's analysis, modeling, data creation, GPS integration, application programming, mobile data collection, web maps, servers and enterprise databases. There's more but we have to go soon. The field is far too big to be an expert in all of the areas but there is substantial crossover for general understanding about how the technology fits together. They all fit together with a common element of geographic data.

Getting started with GIS really depends on what the primary objective is. Do you want to master GIS software to build high quality, professional maps? Or do you prefer to dig into the data with tools and find answers to current day problems? There is always the more casual user who would like to explore data to learn more about their project. Maybe you're the type of person who loves computer systems. Maybe you have computer technology as a primary career path, but are interested in working with providing web mapping services. Perhaps the reader is a computer developer who wants to write GIS applications or build tools within the software. There are plenty of ways to hop in and start playing with open source software, the community is waiting for your entrance.

13.7   That's All Folks

In the words of Porky Pig, "That's all folks!" Hopefully the journey was enjoyable by sharing some history and technical knowledge about GIS that wasn't known before. Although, the spotlight of this journey really is on open source GIS software. All of those experiences and challenges using open source software, and finding ways to plug them into the organization, was really a great time. Unfortunately, all good things do come to an end. But do they? An end? Or is it better said that there is a new beginning down another road with a new journey.

This journey will continue as a freelancer providing GIS services using only open source software. So far it's been great and open source GIS has been used in a wide variety of spatial applications. There are so many interesting projects and people to interact with and help. The work is incredibly interesting and fun. Each project has its own set of unique challenges and specialization. Maybe you can join the mix one day.

13.8   The End?

The end? Nah. This journey never ends so long as other people are going down the open source road. Open source GIS software is here to stay and it keeps getting exponentially better. Thank goodness open source software exists or we'd all be stuck in a commercial GIS software jail cell that limits our thinking and capabilities while simultaneously emptying our pockets. Have a fun and successful journey of your own. We all believe in you. Go get it. Go out and do something great and be sure to share the success stories with others.

Index

Books from Locate Press

Be sure to visit http://locatepress.com for information on new and up-coming titles.

Discover QGIS 3.x

SECOND EDITION: EXPLORE THE LATEST LONG TERM RELEASE (LTR) OF QGIS WITH DISCOVER QGIS 3.x!

Discover QGIS 3.x is a comprehensive up-to-date workbook built for both the classroom and professionals looking to build their skills.

Designed to take advantage of the latest QGIS features, this book will guide you in improving your maps and analysis.

You will find clear learning objectives and a task list at the beginning of each chapter. Of the 31 exercises in this workbook, 7 are new and 8 have seen considerable updates. All exercises are updated to support QGIS 3.26.

The book is a complete resource and includes: lab exercises, challenge exercises, all data, discussion questions, and solutions.

QGIS for Hydrological Applications

SECOND EDITION: RECIPES FOR CATCHMENT HYDROLOGY AND WATER MANAGEMENT.

Now updated - learn even more GIS skills for catchment hydrology and water management with QGIS!

This second edition workbook introduces hydrological topics to professionals in the water sector using state of the art functionality in QGIS. The book is also useful as a beginner's course in GIS concepts, using a problem-based learning approach

Designed to take advantage of the latest QGIS features, this book will guide you in improving your maps and analysis.

Introduction to QGIS

GET STARTED WITH QGIS WITH THIS INTRO-
DUCTION COVERING EVERYTHING NEEDED TO
GET YOU GOING USING FREE AND OPEN SOURCE
GIS SOFTWARE.

This QGIS tutorial, based on the 3.16 LTR version, introduces you to major concepts and
techniques to get you started with viewing
data, analysis, and creating maps and reports.

Building on the first edition, the authors take
you step-by-step through the process of using
the latest map design tools and techniques in QGIS 3. With numerous
new map designs and completely overhauled workflows, this second edition
brings you up to speed with current cartographic technology and trends.

With this book you'll learn about the QGIS interface, creating, analyzing,
and editing vector data, working with raster (image) data, using plugins and
the processing toolbox, and more.

Resources for further help and study and all the data you'll need to follow
along with each chapter are included.

QGIS Map Design - 2nd Edition

LEARN HOW TO USE QGIS 3 TO TAKE YOUR CAR-
TOGRAPHIC PRODUCTS TO THE HIGHEST LEVEL.
QGIS 3.4 opens up exciting new possibilities for
creating beautiful and compelling maps!

Building on the first edition, the authors take
you step-by-step through the process of using
the latest map design tools and techniques in
QGIS 3. With numerous new map designs and
completely overhauled workflows, this second
edition brings you up to speed with current
cartographic technology and trends.

See how QGIS continues to surpass the cartographic capabilities of other
geoware available today with its data-driven overrides, flexible expression
functions, multitudinous color tools, blend modes, and atlasing capabilities.
A prior familiarity with basic QGIS capabilities is assumed. All example data
and project files are included.

Get ready to launch into the next generation of map design!

Leaflet Cookbook

COOK UP DYNAMIC WEB MAPS USING THE
RECIPES IN THE LEAFLET COOKBOOK.

Leaflet Cookbook will guide you in getting
started with Leaflet, the leading open-source
JavaScript library for creating interactive maps.
You'll move swiftly along from the basics to creating interesting and dynamic web maps.

Even if you aren't an HTML/CSS wizard, this
book will get you up to speed in creating dynamic and sophisticated web maps. With sam-
ple code and complete examples, you'll find it easy to create your own maps
in no time.

A download package containing all the code and data used in the book is
available so you can follow along as well as use the code as a starting point
for your own web maps.

The PyQGIS Programmer's Guide

WELCOME TO THE WORLD OF PYQGIS, THE BLENDING OF QGIS AND PYTHON TO EXTEND AND ENHANCE YOUR OPEN SOURCE GIS TOOL-BOX.

With PyQGIS you can write scripts and plug-ins to implement new features and perform automated tasks.

This book is updated to work with the next generation of QGIS—version 3.x. After a brief introduction to Python 3, you'll learn how to understand the QGIS Application Programmer Interface (API), write scripts, and build a plugin.

The book is designed to allow you to work through the examples as you go along. At the end of each chapter you will find a set of exercises you can do to enhance your learning experience.

The PyQGIS Programmer's Guide is compatible with the version 3.0 API released with QGIS 3.x and will work for the entire 3.x series of releases.

pgRouting: A Practical Guide

WHAT IS PGROUTING?

It's a PostgreSQL extension for developing network routing applications and doing graph analysis.

Interested in pgRouting? If so, chances are you already use PostGIS, the spatial extender for the PostgreSQL database management system.

So when you've got PostGIS, why do you need pgRouting? PostGIS is a great tool for molding geometries and doing proximity analysis, however it falls short when your proximity analysis involves constrained paths such as driving along a road or biking along defined paths.

This book will both get you started with pgRouting and guide you into routing, data fixing and costs, as well as using with QGIS and web applications.

Geospatial Power Tools

EVERYONE LOVES POWER TOOLS!

The GDAL and OGR apps are the power tools of the GIS world—best of all, they're free.

The utilities include tools for examining, converting, transforming, building, and analysing data. This book is a collection of the GDAL and OGR documentation, but also includes new content designed to help guide you in using the utilities to solve your current data problems.

Inside you'll find a quick reference for looking up the right syntax and example usage quickly. The book is divided into three parts: *Workflows and examples, GDAL raster utilities,* and *OGR vector utilities.*

Once you get a taste of the power the GDAL/OGR suite provides, you'll wonder how you ever got along without them.

See these books and more at http://locatepress.com

www.ingramcontent.com/pod-product-compliance
Lightning Source LLC
Chambersburg PA
CBHW041639050326
40690CB00027B/5274